SETTLER ECOLOGIES

The Enduring Nature of Settler Colonialism in Kenya

Settler Ecologies tells the story of how settler colonialism becomes memorialized and lives on through ecological relations. Drawing on eight years of research in Laikipia, Kenya, Charis Enns and Brock Bersaglio use immersive methods to reveal how animals and plants can be enrolled in the reproduction of settler colonialism.

The book details how ecological relations have been unmade and remade to enable settler colonialism to endure as a structure in this part of Kenya. It describes five modes of violent ecological transformation used to prolong structures of settler colonialism: eliminating undesired wild species; rewilding landscapes with more desirable species to settler ecologists; selectively repeopling wilderness to create seemingly more inclusive wild spaces and capitalize on biocultural diversity; rescuing injured animals and species at risk of extinction to shore up moral support for settler ecologies; and extending settler ecologies through landscape approaches to conservation that scale wild spaces.

Settler Ecologies serves as a cautionary tale for future conservation agendas in all settler colonies. While urgent action is needed to halt global biodiversity loss, this book underscores the need to continually question whether the types of nature being preserved advance settler colonial structures or create conditions in which ecologies can otherwise be (re)made and flourish.

CHARIS ENNS is a presidential fellow in socio-environmental systems at the Global Development Institute at the University of Manchester.

BROCK BERSAGLIO is an associate professor of environment and development in the International Development Department at the University of Birmingham.

Settler Ecologies

The Enduring Nature of Settler Colonialism in Kenya

CHARIS ENNS AND BROCK BERSAGLIO

UNIVERSITY OF TORONTO PRESS
Toronto Buffalo London

© University of Toronto Press 2024
Toronto Buffalo London utorontopress.com
Printed and bound by CPI Group (UK) Ltd, Croydon, CR0 4YY

ISBN 978-1-4875-5361-6 (paper)   ISBN 978-1-4875-5740-9 (EPUB)
                                 ISBN 978-1-4875-5552-8 (PDF)

Library and Archives Canada Cataloguing in Publication
Title: Settler ecologies : the enduring nature of settler colonialism
   in Kenya / Charis Enns and Brock Bersaglio.
Names: Enns, Charis, author. | Bersaglio, Brock, author.
Description: Includes bibliographical references and index.
Identifiers: Canadiana (print) 2024028254X | Canadiana (ebook)
   20240282566 | ISBN 9781487553616 (paper) | ISBN 9781487557409 (EPUB) |
   ISBN 9781487555528 (PDF)
Subjects: LCSH: Human ecology – Kenya – Laikipia County. |
   LCSH: Settler colonialism – Environmental aspects – Kenya –
   Laikipia County. | LCSH: Wildlife conservation – Kenya –
   Laikipia County. | LCSH: Laikipia County (Kenya) –
   Environmental conditions. | LCSH: Kenya – Colonization.
Classification: LCC GF724 .E56 2024 | DDC 304.2096762/753 – dc23

Cover design: Val Cooke
Cover image: Brock Bersaglio, 2023. An elder walks through land within
a pastoralist community conservancy in Laikipia where invasive *Opuntia*
(prickly pear cactus), introduced by white settlers during the colonial era,
have taken over the landscape.

We wish to acknowledge the land on which the University of Toronto
Press operates. This land is the traditional territory of the Wendat, the
Anishnaabeg, the Haudenosaunee, the Métis, and the Mississaugas of the
Credit First Nation.

This book has been published with the help of a grant from the Federation
for the Humanities and Social Sciences, through the Awards to Scholarly
Publications Program, using funds provided by the Social Sciences and
Humanities Research Council of Canada.

University of Toronto Press acknowledges the financial support of the
Government of Canada, the Canada Council for the Arts, and the Ontario Arts
Council, an agency of the Government of Ontario, for its publishing activities.

 Canada Council for the Arts   Conseil des Arts du Canada

 ONTARIO ARTS COUNCIL
CONSEIL DES ARTS DE L'ONTARIO
an Ontario government agency
un organisme du gouvernement de l'Ontario

Funded by the Government of Canada   Financé par le gouvernement du Canada    Canadä

Contents

Photos

Maps

Foreword

In the foreword to this book, Ramson Karmushu – an activist for inclusive conservation and long-time collaborator – describes what the Laikipia Plateau and its surrounding lowlands and highlands might have looked like prior to colonial settlement. Drawing from insights shared by his grandfather, father, and other elders from his community, Ramson reflects on how the livelihood systems and rangeland management practices of Maasais brought together people, domestic animals, and wild animals in shared landscapes that are now increasingly divided for use by different groups of people and species.

My family is part of the Ilpurko Maasai who have been affected by the two Anglo-Maasai treaties in Kenya during the colonial era. These forced migrations first moved all the Maasai families from the foot of Mount Kenya (Oldonyio Keri) to the northern reserve around Muko-godo Forest in 1904 to create land for the colonial masters, the British, to reward their war soldiers with farming lands – many of which are still occupied by settlers today. All but the Ndorobo Maasai (hunter-gatherers) were subjected to eviction. In the process, huge Maasai grazing fields were made British livestock ranches and crop-farming fields. Knowing the Maasai would eventually demand back their lands for settlement and grazing livestock, all the Maasai were again moved to the southern reserves in 1911. My whole family was therefore moved from Laikipia to Narok in 1911.

My great-grandfather, who was young at the time, was employed at Lord Delamere's farm as a herder to graze the livestock. A settler in Laikipia, Merrick, later went to Delamere to ask for a warrior who he could use as a livestock manager in his farm as the Meru (a farming community) near his property did not know how to herd. Delamere offered one warrior, Ole Maitiko, my great-grandfather. However, Ole Maitiko, who refused to come to Laikipia alone, asked to be given six others, such as Ole Mayiani, Ole Sururu, and Ole Kesier, to name a few.

After they arrived back in Laikipia, Ole Maitiko refused to ever leave again, as he knew it was a good place to herd. He influenced the other warriors and they all married here in Laikipia and stayed here. But he kept visiting his family in Narok until he died very far away. As told by my father, Ole Maitiko eventually died in 1958 and was buried far beyond Isiolo, at a hill known as Olodo Moru, on your way to current Gotu, at a place they called Pina.

The history of my family and that of other Maasai families in Laikipia was heavily shaped by colonialism and white settlement. When I try to imagine a past without colonialism, I think Maasais would have decided to keep their land using a system like group ranches or community land today. Before the time when my family and others were moved, the whole land was not fenced. My ancestors of the Ilaikipiak Maasai used to manage the entire landscape all the way from beyond Archer's Post along the River Ewaso Ng'iro at the border of Samburu and Isiolo to beyond Nanyuki at the foot of Mount Kenya towards the Aberdares (Osupuko Lereko), Nyahururu (Enaiurur), Lake Olbolosat (Oloorpolosat, a place where sacrifices were made), Nakuru (Nakuro), and bare lands connecting to Mau (meaning "twin") Forest and connecting with the southern Maasai at Narok.

During the rainy season, they would go down into the lowlands from Laikipia before slowly making their way back to the area around Nanyuki as the rains finished. Before the rains came again, my ancestors and members of other Maasai families would burn the grasses to control pests, like ticks, and diseases. This way, when the rains fell, new and fresh grass would grow. They did not do this all the time – only when it was necessary, when there was too much dry grass and it had become white and infested with ticks and other pests. They were skilled at controlling fires, so that they did not spread to other areas where grasses still needed to grow and regenerate.

Recalling what I have been taught about those days in the past, I would also expect to see all the wild animals. Some of the animals stayed separate from the livestock, like rhinos, the blood of which was sometimes used as medicine to treat severe illnesses that were not common, such as gonorrhoea, because rhinos consume a lot of herbs. Other animals used to mix a lot with livestock and graze, like the hartebeest. Just like today, no one from my community ever killed or ate that animal. My family were not hunter-gatherers, but even hunter-gatherers would not kill hartebeests. It was a taboo. So again, when I close my eyes to imagine the landscape in the past, I see many hartebeests and they are mixed with cattle. I see other grazers as well. Like today, in the past my ancestors did not eat wild meat in general. The only animals we permitted ourselves to eat were those that behaved like a cow:

buffalos; antelopes; never a zebra; a giraffe, yes; an elephant, never; a rhino, never. However, we would not eat most wild animals. It was not seen as a good thing to eat wild meat.

When I close my eyes and imagine the landscape as my father and grandfather used to describe it, I see bushes and trees with grass growing in between the trees. Parts of Il Ng'wesi, where many in my community now live or graze their animals, still look this way. In the past, Lewa Wildlife Conservancy, next to Il Ng'wesi, used to have lots of trees. You can look at aerial maps and see this in earlier years, such as the 1970s, 1980s, and 1990s. But because fences were constructed to control the movement of wildlife in the name of "conservation," the elephants who could not move out of the electric fences ended up breaking all the trees and destroying them until only grasslands were left. So, when I close my eyes, I picture a landscape that is full of grasses and trees and a good number of livestock and wildlife, as pastoralists were not so impacted by droughts in those days. I see a mix of livestock – cattle, goats, sheep – but cattle were the most prominent and present in greater numbers.

The livestock the Maasai had in those days were indigenous, meaning they were not heavy feeders on the land. The Maasai sheep – which are smaller and have a reddish colour, especially red heads – were more common then, compared to the Merino and black-headed Persian sheep that were introduced by the settler farmers and other sheep that most people now keep. These new breeds of sheep graze and eat a lot, including the grass roots, leaving the land bare and difficult to recover even after the rains. One day I would imagine trying to raise indigenous sheep, like the red-headed ones I have seen between Maralal and Baragoi. Today, though, even my own sheep are exotic, due to their economic value. They are not indigenous.

– Ramson Karmushu, Nanyuki, Kenya, May 2023

Ramson's foreword emphasizes that even though settler colonization profoundly transformed ecologies in the region, other ecologies – which we refer to throughout the book as "ecologies otherwise" – existed prior to settlement and persist or have the potential to re-emerge today. This story continues in the afterword of this book, where Ramson imagines in more detail what a future in which ecologies otherwise are more pronounced might look like. Bookending Settler Ecologies in this way puts a timestamp on settler ecologies, acknowledging that settler ecologies are not the beginning and need not be the end of ecological relations and biodiversity conservation in Laikipia or other settler-colonial contexts.

Acknowledgments

As this book is the combined outcome of both our PhD projects, which started in 2012, we begin by acknowledging those who made the earliest impressions on what would eventually become this book. In the Department of Geography and Planning at the University of Toronto, we extend deep gratitude to Brock's PhD supervisor, Thembela Kepe, supervisory committee, Scott Prudham and Ken MacDonald, and dissertation examiners, Sharlene Mollett and Jim Igoe. We also thank members of the Political Ecology and Economy group, especially Zach Anderson and Lauren Kepkiewicz. Similar thanks are owed to Charis's PhD supervisor, the late Will Coleman, and supervisory committee, Suzan Ilcan and Alex Latta, during her time at the Balsillie School of International Affairs (BSIA) at the University of Waterloo. We are also grateful to Antulio Rosales, Tracy Wagner-Rizvi, Carla Angulo-Pasel, Masaya Llavaneras Blanco, and others from BSIA, including those who attended Monday morning coffee meetings. As we made the transition from PhDs to postdocs, Adam Sneyd (University of Guelph) became a notable source of encouragement and model book writer.

Although the seeds for this book were first sown elsewhere in the world, the book truly took root in Sheffield, England. In 2017, we joined the Department of Geography at the University of Sheffield as a postdoc (Brock) and assistant professor (Charis). Between 2017 and 2019, members of the department and wider community influenced our ongoing research for the book after our PhDs. This includes members of the Geographies of the Global South Research Cluster and Political Ecology Group, such as Adeniyi Asiyanbi, Hannah Dickinson, Laure Joanny, Teresa Lappe-Osthega, George Lordachescu, Judith Krauss, and others acknowledged in more detail below. We are especially thankful to Jenny Pickerill, Frances Cleaver, and Juan Miguel Kanai for the mentorship, opportunities, and support offered to us while in the Department of Geography.

A few people from our time at Sheffield deserve further mention. We are enormously grateful to Dan Brockington, director of the Sheffield Institute for International Development (SIID, now IGSD) at the time, and Rosaleen Duffy, who was leading the BIOSEC project. Dan and Rosaleen not only read and commented on earlier versions of this book, but also showed us that it is possible to create collegial and nurturing spaces within academia. We owe so much to you and your mentorship and will spend the rest of our careers trying to emulate what you created at Sheffield – while wishing we could travel back in time to those days! Through SIID and BIOSEC, we were privileged to be connected to others who would generously offer their mentorship and support during the early days of our careers and this book. Bram Büscher and Tor Benjaminsen deserve shout-outs, as do those connected to their research initiatives, CONVIVA and Greenmentality, respectively.

Part of what made our time at Sheffield so unique and impactful is that we were able to forge friendships with other like-minded early-career political ecologists. Jared Margulies, Francis Massé, Jasper Montana, Elaine (Lan Yin) Hsiao, and Esther Marijnen: we are so grateful for the times we shared in Sheffield. All these people have fed into this book in different ways as it took shape, from discussing research and debating ideas in crowded pubs in Sheffield to giving feedback on draft chapters and manuscripts to eventually organizing book talks. Without such camaraderie, commiseration, and fun, this book would not look like it does.

We also have much gratitude for the wider academic community of political ecologists – and the Political Ecology Network (POLLEN), specifically. Many POLLENators have shaped and sharpened the ideas in this book through debates and discussions and their own work. While there are far too many individuals to name, we are so grateful to be part of this community.

As this book neared completion in 2023, we benefited from being able to receive feedback from newer colleagues in the Global Development Institute at the University of Manchester (Charis) and the International Development Department at the University of Birmingham (Brock) – where each of us now works. Particular thanks are extended to Nic Cheeseman, Jonathan Fisher, and Sam Hickey for their guidance on navigating the publishing process. We presented parts of the book to the More-Than-Human seminar series at the University of Oxford, convened by Jamie Lorimer and Jasper Montana. Bram Büscher, Farhana Sultana, Hannah Dickinson, and their respective colleagues included us in panels at POLLEN, AAG, and RGS. All these spaces provided us with invaluable insights and feedback to consider as we completed revisions on the final version of the book.

This book would not be possible without numerous people outside academia. To start, our families bestowed great privilege upon us as they moved and migrated across the globe. Though we have devoted our lives to understanding, critiquing, and dismantling many of the structures behind this privilege, it is not lost on us that they have made decisions that we continue to benefit from. Over the years, our families have contributed to this book in very practical ways by letting us stay with them or borrow their vehicles, permitting us to repurpose shared family holidays as research opportunities, and extending us immeasurable grace, support, and love as we moved around the world. We have received equal grace and support from friends, of which there are too many to name here. Nevertheless, Brock would like to thank Danny for never saying "no" to a possible adventure. Charis (and Brock) would like to thank Ramsey for her tireless support and sustained interest in our academic projects. In addition to epitomizing friendship, you and Sean have been an incredible sounding board. You are always willing to listen and talk through our ideas and always have challenging and important questions to ask. This is appreciated more than you know.

Many parts of this book were written during the disorienting years of the COVID-19 pandemic, when lockdowns prevented us from seeing most of our friends and family aside from Angela Holzer and Francis Massé. The holidays and weekend getaways spent with Angela and Francis – usually in a remote cottage somewhere in the north of England – and our small pack of dogs were so cherished and vital during such strange times. Although we lost Angela in 2022, she remains close to our hearts, dearly loved, and sorely missed.

To our child, Leighton: Thank you for coming into our lives before this book was finished and for giving us the pleasure of writing you into the story. Our canine companions also supported us in writing this book by insisting we leave the house with them at least once per day when in the throes of writing or responding to reviews.

The research behind this book would not be possible without financial support from the Social Science and Humanities Research Council (on multiple occasions), International Development Research Centre, and British Academy and Leverhulme Trust. Crucially, funding from the University of Manchester also allowed this book to be published open access, and we are so grateful to the university for seeing value in this work and making it as accessible as possible. Thank you to Leonard Luz for designing the maps in the book and to our editors, Jodi Lewchuk and Melissa MacAulay, at University of Toronto Press.

Last, but by no means least, it is difficult to put into words how grateful we are to colleagues and friends in Kenya who we have had the great privilege of working with over the years. Since 2016, we have worked more closely with members of the Indigenous Movement for Peace Advancement and Conflict Transformation (IMPACT) than anyone else. Mali Ole Kaunga, Ramson Karmushu, and Elizabeth Silakan: we do not know where we would be without your friendship and guidance these past eight years. Thank you to Ramson for writing the foreword and afterword for this book and for helping to direct the research behind this book. Neither of us will forget the time we have spent together crisscrossing the vastness of northern Kenya. We are also enormously grateful to Sabina and Silole, and all the women of IMPACT, as well as Naimado, Mema, Tingoi, and David Silakan (formerly of IMPACT). Without doubt, working and walking alongside IMPACT will remain one of the greatest honours of our lives. Outside IMPACT, we owe much to Alex Awiti, who has supported our research since 2014. We are also grateful for our learning from more recent colleagues, including Kennedy Mkutu, Evelyne Atieno Owino, Klerkson Lugusa, Nashipai Seketeti, and Paul Lebeneyio.

SETTLER ECOLOGIES

Introduction

Standing on a hill overlooking Sanga, a Maasai community on the northern edge of Borana Conservancy in Laikipia,[1] Kenya, an elderly man recounts:

> A long time ago, when we were warriors [youth], we used to sit here in this tree and hunt buffalo. At that time, the whole forest was so dense that no people or livestock would walk through it. There were no elephants around these parts then – just the occasional buffalo or leopard. From right here, we only saw elephants from a distance maybe once each year. In fact, it was not until I began taking livestock out on my own that I even saw an elephant up close for the first time.

Now, those living in Sanga encounter elephants on a much more frequent basis. On this day, a herd of about thirty elephants is moving slowly through the community, and we watch them browsing on grass and trees surrounding peoples' homes from where we sit. Electrified fences have been constructed, encircling peoples' homes and a primary school, in a futile attempt to keep the elephants at a safe distance. However, the fences have failed to do their job, and this herd found its way into the community overnight.

The elderly man we are speaking with is standing under an umbrella-shaped tree, known as *Ltepes* or *Acacia tortilis*, depending on who is speaking. Around us, the soil is cracked from drought and the ground

---

1 In this book, we use "Laikipia" loosely in reference to Laikipia County (which includes the Laikipia Plateau) and adjacent areas, such as the Samburu lowlands and Mount Kenya and the Aberdare Range.

is bare of grass, but there are trees and shrubs providing shade. At the bottom of a hill, about 200 metres away from where we stand, there is another high-voltage electric fence. Beyond this fence, the landscape shifts abruptly from the patches of wooded vegetation and shrubs that surround us to vast grasslands. Far off in the distance, tall-standing fever trees can be seen alongside a seasonal riverbed. A tower of reticulated giraffes – a subspecies of giraffe endemic to this area – makes its way towards the river.

The grass-dominated landscape we overlook is the Lewa-Borana Landscape. The landscape is marked with both the ecological aesthetic and the name bestowed upon it by white settlers who arrived in this area over a century ago. The Lewa-Borana Landscape is a conservation area made up of Lewa Wildlife Conservancy and Borana Conservancy, stretching across 380 km². It is situated on the northern edge of Laikipia. Laikipia County spans 9,800 km² and is located on the equator between Mount Kenya, the Aberdare Mountains, and the Rift Valley in north-central Kenya. On a clear day like this in Sanga, you can see the sharp drop-off where the Laikipia Plateau ends, as the Lewa-Borana Landscape bleeds into the vast Samburu lowlands before running up against a range of blueish-grey mountains on the horizon, called *Lenkiyio* or Matthews Range – again, depending on who is speaking.

Around the turn of the twentieth century, Laikipia was declared part of Britain's East Africa Protectorate. Laikipiak Maasais living on this land, along with their livestock, were forcefully evicted to create space for European settlement, causing the death of both pastoralists and cattle (Hughes 2006). Laikipia and the wider central highlands were established as a white reserve with the White Paper of 1923 – excluding Indigenous Peoples and people of colour from owning land within the boundaries of the so-called White Highlands and formalizing a racialized property regime in the process (Bersaglio 2018). Large livestock ranches were started to support the growth of the British East Africa Protectorate's export-orientated free market economy (Pestalozzi 1986) and various settlement schemes were initiated to expand the settler population, formally establishing Laikipia as a space of settler colonialism and white belonging.

It was during this time that livestock ranches, including Lewa and Borana, were founded around the community of Sanga by settlers who were given expropriated land as part of the colonial administration's efforts to increase European immigration to Kenya. During the early decades of colonial settlement, intensive livestock production on this land significantly altered vegetation – for example, reducing woody vegetation and tree cover (Giesen, Giesen, and Giesen 2006) – while hunting decimated wildlife populations. However, when settlers faced crises during the independence era that threatened their landholdings, livelihoods, and futures in Laikipia, they turned to wildlife as a solution. They stopped hunting and removed internal fences on their ranches. Wildlife populations began to move and grow, attracted by the relative safety, vegetation, and presence of permanent natural and human-made water sources on these properties. By the 1990s, settlers across Laikipia were beginning to take advantage of the opportunities created by wildlife by converting their ranches into wildlife conservancies and establishing eco-tourism ventures on their land with the help of international investors and conservation organizations. Today, Laikipia hosts more endangered species than anywhere else in Kenya.

The elderly man we are with in Sanga continues speaking, describing how the growth of wildlife populations in the landscape over the last several decades has affected relationships between people, animals, and plants. He explains:

> The tree we are sitting under now is one that elephants also like to use for shade and to eat from when the season is right. A few months ago, a cow was under this tree and it was killed when a herd of elephants arrived and attacked it. Because there are so few large trees left in this area, there is now always competition for space under the large trees.

The man recalls how, when he kept his livestock in this area during the late 1990s, he welcomed the presence of elephants. "Elephants used to signify safety and peace for herders," he says; "Other threats – people or wildlife – were kept away when elephants grazed nearby." At night, he would set up camp under this very tree for himself and his livestock while elephants would continue to graze or sleep under nearby trees without bothering him or his livestock. Back then, he says, this area was "a densely wooded forest and there were enough trees for all." Today, the forest near Sanga, where elephants frequently visit, looks quite different (see photo 0.1).

The elderly man is not alone in his observations about the changing landscape. Recent studies carried out on the Lewa-Borana Landscape discuss vegetation change in this area over the last several decades. Studies conducted in 2006 and 2016 note declines in tree and shrub vegetation within Lewa's boundaries, with large areas of woodlands having been converted to grasslands since 1962 (Giesen, Giesen, and Giesen 2006; 2017). The studies highlight multiple drivers of this transformation, including ranching practices, fires, and environmental change. Yet the studies conclude that the most notable driver of forest-to-grassland

Photo 0.1. Photo taken of grasslands in Lewa-Borana Landscape while driving on a public road en route to Il Ng'wesi

conversion has been the significant increase in large herbivores such as elephants, rhinos, and giraffes, which cause widespread tree damage (Giesen, Giesen, and Giesen 2017, 4). This damage, which became more pronounced between 2006 and 2016 when wildlife numbers increased dramatically, is linked to the growth of wildlife populations on private conservancies (Giesen, Giesen, and Giesen 2017).

In response to vegetation change on private conservancies, efforts are underway across Laikipia to extend and secure the habitat available for wildlife, with a strong focus on securing more habitat for large, endangered mammals, such as elephants and rhinos. These efforts involve removing fences between properties, annexing new land to expand existing or establish new conservancies, and creating new dispersal areas and migratory corridors so that wildlife can move more easily. A significant step in expanding wildlife areas took place in the Lewa-Borana Landscape in 2015 when the fence separating the two former cattle ranches was removed to form one continuous landscape. Since then, there have been a series of further negotiations with private and community landholders in the area to release more land for wildlife. To the south of Lewa, a tract of land has been converted from farming land to an elephant corridor that allows elephants to move freely between Lewa and Borana, through Kisima Farm and towards Mount Kenya Forest Reserve (GoK 2017). To the south of Borana, a community forest called Ngare Ndare has been opened up to create a wildlife corridor that also links Lewa and Borana to Mount Kenya Forest Reserve. More recently, there have been talks about removing fences to allow wildlife to move unhindered between Il Ng'wesi – the community-owned land on which Sanga is located – and the Lewa-Borana Landscape. This proposal has been subject to heated debate.

Given the history of Laikipia, the removal of fences between settler ranches-turned-wildlife conservancies and community land is not accepted as a strictly ecological intervention. Rather, opening up community land to support growing populations of wildlife invokes a sense that history may be repeating itself – again. For over a century, the ecological relations important to pastoralists have been marginalized as conservationists have altered and reconfigured what types of animals and plants are predominant in the landscape to benefit shifting settler political economies and socialities. For some pastoralists, the most recent proposal of extending the Lewa-Borana Landscape for wildlife represents yet another attempt to recreate and sustain settler ecologies in Laikipia – namely, ecologies that support the continuation of settler colonialism.

Settler Ecologies

Laikipia is renowned as a biodiversity hot spot of global significance. Against the odds given the worsening biodiversity crisis, many wildlife populations in Laikipia appear to be doing reasonably well. Aerial surveys indicate there are likely more large mammals residing in Laikipia than Kenya's Amboseli, Nairobi, and Tsavo East and Tsavo West National Parks combined (LWF 2012). In addition to providing habitat to significant numbers of mammals, Laikipia is incredibly biodiverse. This part of the central highlands contains "95 species of mammals, 540 species of birds, 87 species of amphibians and reptiles, almost a 1000 species of invertebrates and over 700 species of plants" (LWF 2012, 10). Laikipia is home to half of Kenya's black rhinos, the country's second-largest elephant population, and strong numbers of lions, leopards, and cheetahs. Laikipia is also renowned for the protection of endemic species like the Grévy's zebra, the reticulated giraffe, and the Laikipia hartebeest. In fact, the region supports "the second highest abundance of wildlife in East Africa, after the Mara-Serengeti ecosystem, and hosts the highest populations of endangered large mammals in Kenya" (Witt et al. 2020, 517).

Because of the density and diversity of wildlife in more rural parts of Laikipia, alongside a relatively sparse population and large landholdings, Laikipia has become known for specializing in elite and luxury safari experiences. Many tourists will travel to exclusive eco-lodges by helicopter, dropping down in the middle of tens of thousands of acres of open grassland and shrubland. They enjoy lunch overlooking salt licks where Grévy's zebras graze alongside Beisa oryx and they spend their afternoons sipping sauvignon blanc or Tuskers in infinity pools while overlooking watering holes where elephants come to quench their own thirst. Tourists fall asleep in mahogany beds covered with white mosquito nets, waking up in the early morning to birds sounding their alarms to big cats walking near camp. At dawn and dusk, tourists venture out on guided tours to undertake a mix of adventurous forays – many of which one can do only on private land in Kenya: late night game drives, safari on horseback, animal tracking by foot, heli-flying, quad biking, fly fishing, and so on. By the time the helicopter returns for their departure, these tourists may have driven hundreds of miles over the course of their stay without seeing or interacting with anyone beyond those paid to cater to their needs – due to the amount of land set aside for safari tourism and a careful choreographing that ensures the landscape looks and feels as it should.

Today, ecological relations in large parts of Laikipia are entwined with this particular type of wildlife economy, based around elite and luxury safari experiences, that dominates land use on former settler ranches in the region. Safari tourists have strong beliefs about what the landscape that they travel through should look like, based largely on problematic imaginaries of nature in Africa – "images of a 'Wild Eden,' rugged, 'pristine' landscapes, and some of the most charismatic global 'megafauna' (elephants, gorillas, rhinos, etc.) are etched in the mainstream connotations attached to the continent" (Büscher 2011, 84). As John Mbaria and Mordecai Ogada write (2016), to colonizers, Africa is and always has been home to wild and uncontrollable charismatic species that exist to be saved (or shot) by non-Africans. The beliefs, attitudes, desires, and behaviours of safari tourists have real implications for the success of safari destinations and influence what types of ecological relations are allowed and refused in areas used for safari tourism. As one safari guide explained to us, clients that come for such exclusive experiences lose their temper over seeing domestic livestock while on game drives. "I didn't come all this way to look at cows!" a safari guide in Il Ng'wesi recalls one angry tourist shouting at him. Creating and sustaining a safari-based wildlife economy requires producing certain ecological relations that do not deviate from global imaginaries of African nature.

Yet, the ecological relations produced on ranches-turned-wildlife conservancies today are not the same ecological relations produced by settlers in the past. Travel back in time to the mid-twentieth century, and most land in Laikipia was used for large-scale wheat farming and cattle ranching in what was then known as Kenya's White Highlands. Beyond livestock, few, if any, significant populations of large mammals were found on cattle ranches due to over-hunting. This included hunting by settlers, who saw wildlife as either pests or trophies and hunted them prodigiously until a nationwide hunting ban in the 1970s (Steinhart 2006). Go back further to the 1920s and 1930s and the white settlers who had only recently acquired land in Laikipia were actively culling wild plant and animal species, replacing them with European imports or hybrids and clearing the land of shrubs and trees for pastures and farms. All of this was made possible by colonial land alienation policies, the violent eviction of Maasais and other groups to make way for white settlement, and, eventually, policies of racial segregation that formalized the White Highlands. Travel back even further in time to the period before colonial settlement and ecological relations in Laikipia would have looked different yet again, with more diverse vegetation – bearing indigenous fruits, herbs, and roots – spread across what is now grassland and freedom of movement for indigenous livestock species and wild fauna across the landscape (Lolwerikoi 2010). Although conservation efforts in Laikipia are acclaimed for restoring a primordial state of nature, the contemporary conservation landscape is better understood as an imagined and new iteration of nature.

Settler Ecologies examines how settler colonialism has impacted and endures through ecological relations, including relationships between humans, animals, and plants. Focusing on Laikipia as a case study, we show how the continued unmaking and remaking of ecological relations by those we describe as settler ecologists works to secure past settler-colonial advances and a future for settler colonialism. We introduce five modes of violent ecological change enacted by settler ecologists at different points in time to manage biodiversity and rearrange ecological relations in ways that prop up and prolong settler colonialism. These are *eliminating* undesirable wild species to create productive landscapes; *rewilding* landscapes with more desirable species to make them productive once degraded; selectively *repeopling* wilderness to create seemingly inclusive wild spaces and capitalize on biocultural diversity;

rescuing species at risk of extinction, and orphaned and injured animals, to shore up moral support for settler ecologies; and extending the range of settler ecologies by *scaling* wild spaces.

We build on a well-established body of political ecology literature that problematizes the ways conservation is marked by coloniality and perpetuates historical and neocolonial violence (Neumann 1998, 2001; Singh and van Houtum 2002; Ramutsindela 2004; Brockington and Igoe 2006; Kepe 2009; Akama, Maingi, and Camargo 2011; Mbaria and Ogada 2016; Griffiths et al. 2023). This work has shown how the conservation sector has evicted existing and rightful land users before fortifying and securing lands to create protected areas for wildlife in many post- and settler-colonial contexts (Brockington 2002). This, in turn, has produced exclusionary, racialized, and segregated landscapes meant to appeal to (neo)colonial imaginaries of pristine and peopleless nature in Africa (Neumann 1998, 2001; Brockington and Igoe 2006; Igoe 2017). There is also a large body of literature that shows how dominant norms and practices in conservation today continue to be marked by coloniality[2] (Singh and van Houtum 2002; Ramutsindela 2004; McGregor 2005; Kepe 2009; Koot, Büscher, and Thakholi 2022; Akama, Maingi, and Camargo 2011; Adams and Mulligan 2012; Mbaria and Ogada 2016; Thakholi 2021), including the notion that nature needs to be separated from humans to be effectively conserved.

Existing work on the political ecology of conservation has tended to be anchored by a degree of human exceptionalism, with less attention paid to how specific animal and plant species, as well as more-than-human entanglements, have been subjected to and enrolled in enactments of injustice (Srinivasan and Kasturirangan 2016; Margulies and Bersaglio 2018). In *Settler Ecologies*, we reveal how ecological relations – defined as "human and nonhuman living beings (plants, animals, persons, insects), nonliving beings and entities (spirits, elements), and collectives (e.g., forests, watersheds)" (Whyte 2018a, 126) – that settler ecologists work to create and conserve can serve as conduits for settler colonialism, just as property regimes and labour relations in conservation often do. We argue that settler ecologists' affinity for certain species and disdain for others has informed policies and practices of conservation for over a century and, accordingly, has shaped the social and

---

2 Coloniality can be understood as "longstanding patterns of power that emerged as a result of colonialism, but that define culture, labour, intersubjective relations, and knowledge production well beyond the strict limits of colonial administrations" (Maldonado-Torres 2007, 243).

environmental justice implications of conservation. Our contribution to this work draws the concerns of environmental history (Crosby 1989; Griffiths and Robin 1997; Beinart and Hughes 2007) into the present, showing how certain species and specific ecological relations continue to be used in support of present-day settler-colonial projects (Todd 2014, 2017; Blair 2017; Taschereau Mamers 2019; Dicenta 2023).[3]

The remainder of this chapter is structured as follows. The next two sections introduce the conceptual framing of the book, linking our discussions about settler ecologies and settler ecologists to wider literature theorizing settler colonialism. Then, we discuss the methods and research that informed this book. Next, we reflect on our own positionality in relation to settler ecologies. We finish this chapter by outlining the structure of the book.

Settler Colonialism and Ecological Relations

What differentiates colonial and settler-colonial forms of conquest is the colonizers' relationship to the colonized territory and its inhabitants (Wolfe 1999). Colonialism involves the extraction of materials and exploitative or indentured labour to metropoles – the homelands of colonial powers. In contrast, settler colonialism is undertaken through the ongoing occupation of colonized land by settler populations: settler colonialists arrive with the intention of remaining permanently (Velednitsky, Hughes, and Machold 2020). As Patrick Wolfe argues, what makes settlers different is that they "come to stay," making settler colonialism an enduring structure, rather than a single historical event (Wolfe 2006, 388).

Acquiring territory is the necessary first step of all settler-colonial projects. Oftentimes, shoring up access to territory was and is made possible by *terra nullius* – claims that lands to be occupied are empty, unused, and awaiting settlement. In historical writing by settlers, landscapes subjected to colonial settlement were routinely described as places of waste or emptiness, waiting to be brought to life by "the brilliance and ingenuity of rugged and ambitious arrivistes" (Hugill 2017, 6). Such narratives were at play when settlers arrived in Kenya. Following an 1883 expedition in Kenya funded by the Royal Geographic Society, Scottish explorer Joseph Thomson wrote, "The greater part of

---

3 After this book went into production, Irus Braverman published *Settling Nature*, documenting "nature's power in the hands of the Zionist settler state" (2023, 1). Focusing on Palestine-Israel, Braverman's work offers another example of the power of settler ecologies in a different context.

Lykipia – and that the richer portion – is quite uninhabited, owing, in a great degree, to the decimation of the Masai of that part" (1887, 238). The (mis)representation of Laikipia as a vast space lying idle and empty was used as justification for white settlement (Huxley 1948, 1953; Collett 1987).

In reality, the lands that settlers wish to settle are rarely if ever empty to begin with. Instead, Indigenous presence is denied rather than being an accurate reflection of reality (Veracini 2008; Miller et al. 2010; Dlamini 2020). Most settler-colonial projects require that expropriated territory be cleared to create space for settler occupation and use. As Alan Lawson and Anna Johnston write, settler colonialism proceeds by first eliminating "those who, since time immemorial, have lived on, defined themselves in terms of, and taken responsibility for that land" (2000, 28). Settlers have used a wide range of tactics to empty land so that it can be settled, including indirect and direct forms of violence to eliminate Indigenous Peoples from the land base, ranging from entrenching property regimes in which Indigenous populations are denied land title to forcibly moving Indigenous Peoples to reserves to enforce the spatial segregation of races (Tomiak 2016, as cited by Hugill 2017, 6). Such tactics place "both the power of the state and the control of economic resources, particularly land and other means of production ... directly in the hands of settlers, who [control] and [use] state power discriminately against the native to their own advantage" (Odukoya 2018, 178).

With this in mind, settler colonialism is often described as having two defining characteristics. First, settler colonialism is guided by a "logic of elimination" as it "destroys to replace" (Wolfe 2006, 388; also see Sayegh 1965). This includes the elimination of Indigenous Peoples by settlers in order to establish themselves on native territory (Wolfe 2006, 388), as well as the elimination of existing polities as settler colonialists attempt to exercise sovereignty over the territories they have claimed (Saito 2014, 22). Second, settler colonialism is an ongoing and enduring structure rather than an event relegated to the past. Unlike other forms of colonialism, "the colonising community remains behind after the end of empire, to capitalize on the unequal social populations with the colonized population" (Mamdani 1996, 14). As Wolfe argues, what makes settlers is that they "come to stay" and, as a result, settler colonialism is "a structure not an event" (Wolfe 2006, 388).

A key contribution we make in this book is to multispecies understandings of settler colonialism. Environmental historians have already shown how the introduction of European animals, plants, and diseases into Indigenous territories was fundamental to colonization (Crosby 1989; Griffiths and Robin 1997; Beinart 2003). More recently, research has begun to focus on how settler colonialism manifests through more-than-human beings, entities, and entanglements in other contexts (Todd 2014, 2017, 2022; Blair 2017; Taschereau Mamers 2019; Gillespie and Narayanan 2020; Dicenta 2023; see also Lenzner et al. 2022). This book adds to this work. It shows how both of the defining characteristics of settler colonialism outlined above – its logic of elimination and its endurance – are extended through the more-than-human world, resulting in the ongoing erasure and replacement of existing ecological relations with those of use and value to settler colonialism.

What is to be gained from a multispecies analysis of settler colonialism in Laikipia? As Zoe Todd writes, "Understanding how settler colonialism structures itself in the lands, waters, and atmospheres that it invades gives us the power to refract its efforts and assert something liberatory in its place (2017, para. 3). Multispecies perspectives are therefore essential to "revealing, supporting and pursuing radically different ecologies that are more inclusive and just" (Mabele et al. 2021), and by extension societies too. By revealing how relations between humans, animals, plants, and other entities in landscapes can be seized and altered to serve settler-colonial interests and prop up structures of settler colonialism, this book represents one step and approach to pursuing these different ecologies, including those that once existed, that persist in defiance of settler colonialism, or that could exist in the future.

Settler Ecologists

In addition to demonstrating how settler-colonial projects are advanced through ecological relations, part of what we aim to do in this book is to unpack who is involved in creating and sustaining these ecologies. Our definition of "settler ecologist" is informed by ongoing debates in settler colonial studies about what it means to be a settler. Some scholars argue that because the primary motive for settler colonialism is access to territory, neither race nor migration by choice matter: settlers are all those who participate in settler-colonial structures motivated to eliminate Indigenous People (Wolfe 2006; Lawrence and Dua 2005). As Tuck and Yang argue, every non-Indigenous person in the settler state is "invited to be a settler in some scenarios, given the appropriate investments in whiteness" (2012, 35) – investments that are generally made by acquiring Indigenous territory. From this perspective, regardless of why someone arrives in a settler colony, they could be seen as settlers as could their descendants if they chose to stay.

There has been pushback against broad-sweeping conceptualizations of who counts as a settler, with many rightly questioning whether involuntary migrants – such as formerly enslaved peoples, refugees, or those who are undocumented – are settlers (Day 2015). Those questioning broader definitions often argue against collapsing all non-Indigenous racialized groups into the subject position of "the settler" because the exploitation of racialized people – Indigenous or otherwise – lies at the core of settler colonialism (Day 2015), resulting in different but lasting experiences of marginalization, violence, and landlessness. In light of these arguments, terms such as "migrant" or "arrivant" have been proposed to distinguish groups that were forced to move to colonies and never sought to establish their own sovereignty from settlers even if they did stay permanently (Veracini 2010; Byrd 2011). This ongoing conversation suggests that it is important to be aware of all non-Indigenous peoples' positions in structures of settler colonialism while holding on to the knowledge that these positions are acutely differentiated by history and context (for more, see Sharma and Wright 2008; Saranillio 2013; Pulido 2017; Upadhyay 2019).

Given this debate, we do not propose an overarching definition of "settler ecologist" or a typology of who is and who is not a settler ecologist in this book. Instead, we emphasize the need to understand how context and history shape different individuals' and groups' roles within, and relationships with, settler ecologies. Settler ecologists include, but cannot be reduced to, people of European descent who arrived in Laikipia during the colonial era. A much wider cast of characters has been and also remains involved in creating settler ecologies, and this cast has changed at different points in time. When we speak of settler ecologists in this book, we refer to all those who consciously and unconsciously participate in the elimination and replacement of existing ecological relations with those that enable structures of settler colonialism to endure. At the same time, and as will be evident in the chapters that follow, we are careful to recognize that these characters hold different degrees of power, agency, and responsibility in producing ecological relations that sustain settler colonialism as a result of how they have been differentially enrolled in settler-colonial projects.

Each empirical chapter of this book identifies different actors that have been involved in ecological transformations that support settler colonialism at different points in time. White landowners who arrived to settle Kenya over the last century or so, along with their descendants, are key characters throughout the entire story. The population of white people living in Kenya today that descended from settler families numbers between 3,000 to 5,000 (McIntosh 2017, 662). However, according to the Kenya National Bureau of Statistics (2019), this population is much larger: there were 69,621 Europeans residing in Kenya, of which 42,898 were Kenyan citizens (KNBS 2019). Many of these people are of British origin, but they have also come from other European countries, like Italy and Greece, and other settler colonies, like South Africa. Kenya's white constituency amounts to around 1 per cent of its total population. Until recently, many white Kenyans lived in affluent suburbs of Nairobi, such as Karen or Gigiri, but there has been a notable growth in the proportion that reside in Laikipia in recent years. Today, some forty white settler families own roughly 1 million acres of land in Laikipia, controlling nearly 40 per cent of the total land area in the county (McIntosh 2017, 663; see also Letai 2015). As chapters 1 and 2 detail, this land has been used for a mix of activities since the early 1900s, which have served as important industries for advancing settlers' ecological sensibilities and wider interests in the landscape, including farming, ranching, hunting, and, most recently, conservation and tourism.

Acting alongside white landowners are several other groups that have become intimately involved in rearranging and managing ecological relations in ways that sustain settler colonialism. These include international conservation organizations; national conservation organizations; private and public security actors; conservation researchers; national environmental management and wildlife authorities; philanthropists; and conservation investors. Large international conservation organizations, like Fauna & Flora International (FFI), that were established by colonial powers provide fairly clear-cut examples of settler ecologists beyond white settlers – particularly as such organizations have begun buying up large tracts of land that they manage using tactics and strategies resembling militarized and fortress conservation. In contrast, individuals doing conservation research may seem less obvious examples of settler ecologists upon first glance. However, as we show in several of the chapters that follow, such individuals can play remarkably important roles in sustaining settler colonialism by producing research that supports the ongoing erasure and replacement of Indigenous ecological relations with those of use or value to settler-colonial projects.

As mentioned above, we recognize that different types of settler ecologists operate with different capacities, motivations, and agency – often as a result of the different ways they are positioned in relation to settler-colonial structures. For example, white landowners exert strong influence over what types of animals and plants are predominant on their private land and, consequently, in the wider landscape. At the same time, wildlife authorities, conservancy managers, and armed private security forces play a key role in making, promoting, and upholding the landscape aesthetic that settler ecologists work to create. We neither fully include nor fully exclude these actors – some of whom are black and Indigenous Kenyans – in our discussions about settler ecologists. Instead, we recognize power differentials while still drawing attention to how these actors' labour is made useful or complicit in processes that advance settler-colonial projects (Fung 2021; Upadhyay 2019). In such instances, the label of "settler ecologists" is used to show how settler colonialism works through diverse peoples' bodies, minds, and labour – without denying that these people may experience both some benefits through settler-colonial power relations (e.g., relatively good wages, job security, and employment benefits) as well as colonial violence and structural racism.

Ultimately, our use of the term "settler ecologist" is meant to reveal the assemblage of actors involved in the erasure of existing ecologies and their replacement with those valued and reified by settler colonialism. Although white Kenyans are both central to and firmly involved in this process, many other actors are also implicated at different points in time and in a variety of spaces. As we show throughout the book, even individuals, institutions, and organizations acclaimed for advancing more progressive approaches to conservation are still often complicit in using ecology to keep past settler-colonial advancements in place.

Researching Settler Ecologies

The argument we develop in this book is built around a case study of Laikipia (see map 1.1). Today, Laikipia is often heralded by conservationists as a successful example of a flourishing biodiversity conservation sector in East Africa, partly because the region offers an innovative model for decentralized conservation initiatives on private property and other land not under direct control of the state. Protecting wildlife outside national parks and reserves is a priority for conservationists in countries like Kenya where as much as 75 per cent of wildlife lives outside the boundaries of state-owned protected areas. Laikipia's well-established private property regime – which is a product of colonial settlement – is dominated by large-scale wildlife-compatible land uses. As Mbaria and Ogada (2016) have argued, the region's biodiversity conservation sector has not only preserved colonial land ownership structures, but it has also sustained coloniality as other conservation actors – such as foreign investors and conservation scientists – have been able to build lodges, erect fences, and conduct wildlife research with limited oversight by the government.

With this in mind, Laikipia offers an ideal case study for our research, as there is already an established relationship between colonialism and the biodiversity conservation sector in this region. This provided us with a firm foundation upon which to build our argument that settler colonialism is embedded in the actual ecologies of place and vice versa. Although the methodological approach we used to do this research was in some ways similar to approaches used in existing political ecology work on conservation in postcolonial and settler-colonial contexts – for example, we interviewed conservationists and communities affected by conservation – we were also intentional in thinking about methods that would bring ecological relations into better focus. Much like how tracking animals involves looking for signs left in the landscape that can be used to identify species and attributes of individuals and deduce information about their behaviours and movements, the methodology we devised for this research involved looking for signs across multiple sites, scales, and species of the ecological relations that exist in Laikipia, including signs of how these relations have changed over time. In order to bring into view multispecies encounters and entanglements and draw attention "not only to multiple species but also to how they shape one another" (Collard 2020, 27), our methodology has been multi-sited, multi-scalar, and multispecies.

Between 2015 and 2023, we spent over a year and a half in Laikipia. Some of this time was spent in Nanyuki, Laikipia's largest town, and some was spent living out of town near or on conservancies. Early in our research, we stayed in a housing development that was largely occupied by the people we describe as settler ecologists. The house backed directly onto a large, private wildlife conservancy and bushbucks, impalas, and zebras roamed freely on the grounds of the estate. We also spent periods of time in tents and bandas on privately owned land and community land in and near private conservancies (see photo 0.2). During our day-to-day life in these spaces, we were able to observe and participate in both settler ecologies and ecologies otherwise. We held our meetings, worked from our laptops, and spent our leisure time in the same spaces where settler ecologists also go to grab coffee, drink beer, or swim. Over time, we have become increasingly familiar with the places that settler ecologists frequent, the conservation practices they undertake, and the discourses and thought patterns that shape their practices. By accessing these spaces, we refined our understanding of how settler ecologies in Laikipia have been made, unmade, and remade.

As part of our research, we participated in around twenty safaris at ranches and conservancies managed by white Kenyans, on conservancies owned and operated by international organizations (such as Fauna & Flora International and The Nature Conservancy), and on community conservancies. We spoke with owners, employees, and guests at conservancies about our research while also listening to conversations that took place around us about the history and nature of ecological relations in the area. We hung around restaurants, dining halls, lounges, pool areas, and offices within safari camps. When possible,

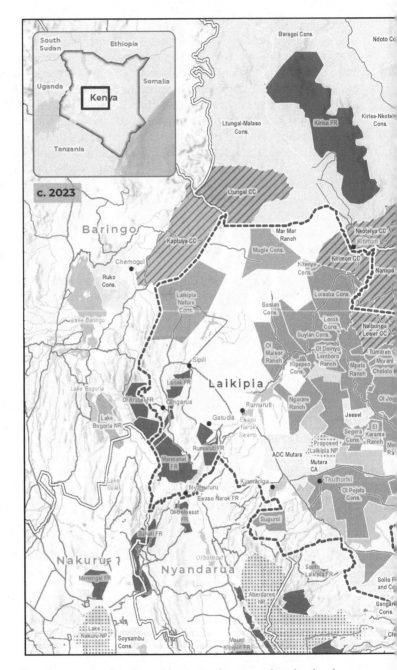

Map 1.1. Map of Laikipia Plateau and surrounding lowlands

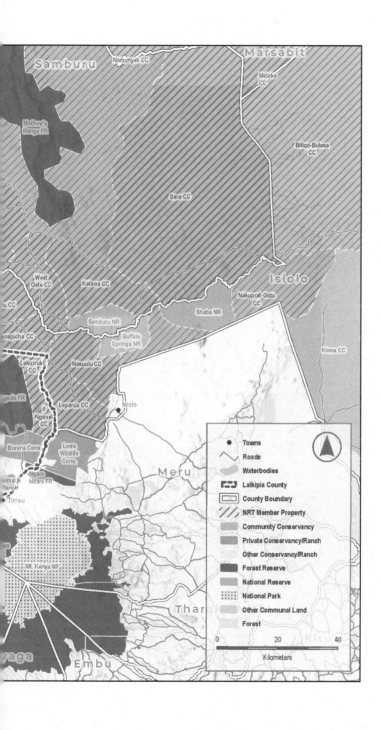

Photo 0.2. Author (Charis) watching zebras in camp on a newly rewilded property in Laikipia

we also engaged in game viewing activities, spending time in 4×4 safari vehicles, on horseback, and on guided walks. We also visited cultural performances and curio shops geared towards safari tourists. We spoke with owners, employees, and guests at conservancies about our research while also listening to conversations that took place around us about the history and nature of ecological relations in the area. Through these conversations and interactions, we collected observational data that led us to identify recurring themes in the making and unmaking of settler ecologies that we examine in this book: eliminating, rewilding, repeopling, rescuing, and scaling.

We conducted over 150 interviews to triangulate our observations and informal conversations. Some of these interviews involved studying up – speaking to both settlers and non-settlers that hold positions of power in Laikipia's conservation sector – given our interest in understanding how a landscape that sustains settler colonialism has been produced and reproduced over time. Such interviews were often carried out in formal spaces where decisions about conservation are made, such as conservancy headquarters, conservation organization offices, and the offices of community, local, and national government representatives. Notably, these spaces and the powerful people within them were far easier to access when we began our research in 2015, before Laikipia's conservation sector was subjected to the growing level of scrutiny it is now. We also conducted interviews with people who experience, navigate, and contest settler ecologies on a daily basis, including those that work in the safari sector and those that live on the fringes of protected areas. These interviews tended to be done in settings beyond the earshot and sightline of settler ecologists – for example, in particular bars and cafes, *manyatta* and pastures, houses, taxies, and even over virtual fora, such as WhatsApp.

Beyond participant observation and interviews, we gathered and analysed articles, (auto)biographies, books, images, maps, policies, reports, and videos produced by a range of actors, including civil society and international organizations, colonial administrators, explorers and settlers, governments agencies, safari companies, scholars and researchers, and tourists. We made use of social media accounts and websites belonging to international organizations, safari companies, wildlife conservancies, and individuals to collect and analyse informational and promotional materials. In some cases, participants provided documents to us – produced by their organizations or even taken from their personal libraries. We also collected historical information from documents in Kenya's National Archives (KNA). We spent seven days in KNA reading and analysing documents to further contextualize the making of Laikipia's settler ecologies in 2015, and then another three days during a follow-up visit in 2017.

Finally, over the course of this book's development, we also became involved in other research projects with civil society researchers, which have provided new avenues to understanding ecological relations in Laikipia. One project we worked on between 2017 and 2019 involved documenting oral histories of pastoralist communities in the lowland areas surrounding Laikipia on strategies for coexisting with wildlife, as well as how relations between people, livestock, and wildlife have changed over the past few generations. For this project, we conducted twelve walking interviews with fifteen participants in Il Ng'wesi where our conversations were enlivened by following animal tracks, learning place names, and sampling herbs. More recently, we have become involved in a larger program of work funded through the Inclusive Conservation Initiative, which seeks to evidence pastoralists' contributions to biodiversity conservation and strengthen conservation initiatives led by pastoralist communities. This work has begun by working with elders from Il Ng'wesi and other communities across Laikipia to document and photograph plant and animal species of value to pastoralist communities. These different projects have allowed us to spend extended periods of time in group ranches, community conservancies, private conservancies, and other wildlife areas. Alongside the collaborations behind them, these projects have also deepened and extended our understanding of the ecologies of the region by providing us opportunities to observe ecological relations from different viewpoints: for example, what we have learnt about settler ecologies while watching elephants graze from the safety of a safari vehicle full of tourists is very different from what we have learnt about settler ecologies while watching elephants in the forest while walking with elders.

We have elected to anonymize many of the quotes and stories we share in this book. Laikipia's conservation sector is always growing, but remains relatively tight-knit. People within the sector often attempt to remain on good terms with one another – even if they hold differing opinions about how conservation should be done. We know that some people we spoke to that were highly critical of settler-led conservation continue to work closely with or for settlers – sometimes trying to transform the sector from the inside and sometimes because it is the best option for more practical reasons, like pay or quality of life. To prevent heightening tensions between different conservation actors, we use pseudonyms when referring to specific individuals throughout the book, although we are as descriptive as we can be without jeopardizing the anonymity and confidentiality of our sources. In a few instances, we have created composites out of insights garnered from more than one respondent or situation to further protect identities. We also tend not to use the actual names of wildlife ranches and conservancies for the most part. In cases where we do refer specifically to individuals, conservancies, or organizations by name, it is because the information we are sharing is publicly available – for example, through news media, published reports, social media, or websites. For ease of reading, the book is also written using "we," although there were some situations where only one of the authors was present.

We comment on the generalizability of this study at greater length in the conclusion. However, for now, what we will say is that despite the known shortcomings of case study research – such as the fact that findings from a single case study are not always broadly generalizable – there are benefits to this approach to inquiry when it comes to developing and articulating new concepts and theories (McLeod 2010). Laikipia offers a useful case study for making the concept of settler ecologies visible and for identifying different approaches used by settler ecologists to keep past settler-colonial advancements in place. Thus, the exploratory and explanatory work done in this book is meant to encourage thought and conversation about how settler ecologies have been produced and reproduced in other settler-colonial contexts, as well as varying manifestations of settler ecologies around the world.

Our Position in Settler Ecologies

We both occupy somewhat unique positions within settler ecologies, and this positionality provides the "political point of departure" (Knight and Deng 2016, 106) for this book. As Stuart Hall writes, "We all write and speak from a particular place and time, from a history and a culture which is specific. What we say is always 'in context,' *positioned*" (Hall 1990, 222). We recognize that research is never an objective undertaking (Haraway 1990). Positionality "affects both substantive and practical aspects of the research process" (Carling, Erdal, and Ezzati 2014, 37) and writing is "always political, interpretative and performative" (Denzin 2019). In this section, we reflect on how our own positionality has informed our research, our writing, and our political aim with this book, which is to challenge the inevitability of settler ecologies and hold space for alternative and Indigenous-led ecological futures.

In thinking about our own positionality, we find thinking on "the space between" instructive (Dwyer and Buckle 2009). This work acknowledges that the strict insider/outsider dichotomy is not always useful in defining researchers' positions in relation to their participants, as researchers often fall somewhere between being complete insiders and outsiders. We occupy multiple in-between spaces that have a significant influence on our research philosophy and style: we are settlers who have lived in Kenya, but we are not part of Kenya's settler community; we research conservation, but we are not ecologists; we have a deep affinity for wildlife and nature, but we are no longer enamoured with settler ecologies and wilderness imaginaries. Most importantly, settler colonialism shaped the contours of our lived experiences and knowledge growing up (see D'Arcangelis 2018, 340); however, we have since slowly begun to unlearn our settler-colonial knowledge, values, and politics – largely by being graciously invited into spaces where we have been guided and requested to do so.

Although we now live in the UK, we are both Canadian citizens of European descent – meaning that we are settlers in the Canadian context. Yet we both have families who settled in different countries in Africa during the post-independence era. Charis's grandparents moved from Canada to Kenya in the 1970s, where her grandfather worked at a college started by missionaries. Charis's mother grew up in Kenya, attending a boarding school in Kijabe, on the edge of the Rift Valley, that was established in 1910 to educate settler children. Brock's family relocated from Canada to the Copperbelt Province of Zambia in the 1990s. His parents were recruited by family friends to manage a childcare program in the mining town of Kitwe. In the early 2000s, Brock was sent to complete secondary school at the same settler boarding school that Charis's mother attended.

Charis was born in Canada, but routinely travelled to eastern and southern Africa growing up to visit family and friends. Trips to Kenya always involved going on safari in places like Maasai Mara and visiting tourist attractions like David Sheldrick Wildlife Trust Elephant Nursery in Nairobi National Park. During her early teenage years, Charis hoped to become a large-animal veterinarian. Brock, too, spent his school holidays on safari with his family in places like Livingstone and South Luangwa National Park. He dreamed of becoming a wildlife artist and would spend his free time studying and sketching animals he encountered on safari. After arriving at boarding school in Kenya, Brock and his friends would spend weekends and midterm breaks exploring Kijabe Forest, Mount Longonot, and Hell's Gate National Park. Brock would eventually return to Canada with the hope of obtaining a university degree in art and pursuing his dream of becoming a wildlife artist.

During our university years, we spent time independently in Canada, Kenya, and Zambia, before moving to Tanzania together for work after finishing our masters degrees. In 2012, we returned to Canada to begin doctoral studies, hoping this would help improve our long-term employability in East Africa. Charis's doctoral research looked at natural resource governance and land politics in Kenya, while Brock began research on private conservation and race in Kenya. It was during these formative years that we began to question our own knowledge, politics, and plans for the future. We became more familiar with the violence and trauma caused by settler colonialism in Canada and began to make connections with settler occupation in parts of Kenya. At this time, we also became involved in work in Kenya that allowed us to be physically present and to learn experientially in spaces of Indigenous-led advocacy. These brief personal accounts offer insight into how our life trajectories, identities, and notions of belonging have positioned us in the in-between spaces that shape our research.

Our in-betweenness in Kenya leaves us well positioned to research settler ecologies. By making settler ecologies the central referent of our research, we are able to sidestep "invasive inquiry" into "the subjectivity of the Other" (Tuck and Yang 2014, 815) – a tendency of geographers and political ecologists to turn the research gaze towards Indigenous communities (Tuck and Yang 2014, 811) – and instead focus our substantive research attention on a community with whom we share similar, but not identical, privileges. Through our life experiences, we have been endowed with implicit and intimate knowledge of settler lifeworlds. This knowledge has been gained through a life of immersion in various settler-colonial contexts, where we have been indoctrinated with the environmental values of settlers and have participated in cultural performances – like safari – that are fundamental to upholding settler ecologies. Through this immersion, we embody an air of familiarity and demonstrate a unique set of cultural competencies that are useful for doing research on the topics addressed in this book. When we arrive at a conservancy driving a small 4×4, dressed in linen and khaki – baby in tow – and Brock comments about an interesting species of bird or type of wildlife spoor observed on our drive, Brock's accent often leads to a fairly predictable set of questions: "How long have you lived in Kenya?" "Where did you go to school?" "Do you know this person or that person?" "Did you play rugby?" Presumptions of shared connections, experiences, and interests guide conversations in directions they might not travel to as naturally otherwise.

Still, despite long family histories in Kenya, we are not part of Kenya's settler community. We are very much outsiders, with our families being part of Kenya's expatriate[4] community instead. Theoretically, what differentiates these groups is that expatriates tend to sustain their own ethnic identities and nationalities while white Kenyan settlers are less likely to identify strongly as belonging to another country (Bickers 2010). In reality, there is something much more ineffable about the dividing lines between Kenya's settlers and expatriates, which may give away upon first impression to settlers that we are not them. Our outsider status means that we may not have been granted the same level of trust and openness that an insider would have been during our research. It also means that those inside this community may feel as though this book lacks true understanding of their history, culture, and ecological aspirations.

At the same time, our outsider status has advantages. Our arm's-length distance from Kenya's white settler community has likely made it easier for us to spend time in spaces and forge relationships with people who do not look as fondly upon Laikipia's settler-dominated conservation sector. It also allows us to avoid the "loyalty tugs" (Coghlan 2007) that one may face when trying to write honestly about their own community. More fundamentally, this distance may make it easier for us to see the links between Laikipia's ecology and settler presence, allowing us to imagine and value ecological futures outside of settler-colonial control. As Joudah writes, "settler permanence is often discussed or accepted as a political inevitability; embedded in this assumption is also a deep spatial permanence" (2020). This assumption of spatial permanence has become entwined with settler ecologies, as settlers have positioned themselves and their landholdings as necessary to securing populations of critically endangered species in Laikipia. Our position of in-betweenness makes it easier for us to pursue questions on alternative and Indigenous-led ecological futures – where the coexistence of critically endangered species and Indigenous Peoples' polities and socialities is possible.

With all this in mind, we do acknowledge that the choice we make to centre settlers and their ecologies in our research may raise important questions. Most pointedly, some would argue that our analysis risks (re)centring settler ecologies in stories about the past, present, and

---

4 We acknowledge important discussions about structural racism and white supremacy associated with the term "expatriate." For example, see Benton (2016).

future of Laikipia while relegating ecologies otherwise to the sidelines. We deal with this concern at greater length in the conclusion; however, our feeling is that by narrating the genealogy of Laikipia's ecological present we are able to show that this ecology is human-made, rather than natural, and tends to serve settler-colonial interests. This, in turn, unsettles the perceived permanence of existing ecological relations and presents Laikipia's ecological future as open for negotiation in the pursuit of a more just multispecies future.

Making Settler Ecologies

This book is structured around the unmaking of existing ecologies and remaking of new ecologies on expropriated land in Laikipia and surrounding areas since the earliest days of colonization in eastern Africa. Each chapter introduces a different approach to ecological transformation used by settler ecologists to produce and reproduce ecological relations that secure past settler-colonial advancements: (1) eliminating, (2) rewilding, (3) repeopling, (4) rescuing, and (5) scaling. Collectively, the chapters describe different actions taken by settler ecologists on the plateau at various points in time to (re)produce structures of settler colonialism, drawing attention to the specific species and unique ecological relations enrolled in and subjected to these ecological transformations. Each chapter also analyses the unique assemblage of settler ecologists involved in altering and reconfiguring ecologies and unpacks the implications for ecologies otherwise.

Chapter 1, "Eliminating," examines how colonial settlers transformed the landscapes of Laikipia between their arrival in the late 1800s and Kenya's independence in the early 1960s. Recognizing the eradication of indigeneity is fundamental to settler colonialism (Wolfe 1999, 2006), we show how the violence of settlement extended to the ecological realm as settlers unmade existing ecological relations and replaced them with new ones that better served their interests and sensibilities. Specifically, we highlight three ways settlers attempted to eliminate species and ecologies predominant in Laikipia upon their arrival. They cleared and culled occupied land of select fauna, flora, and undesired landscape features; hunted species for sport and as part of their livelihood portfolios; and implanted desired, and often imported, species and genes to weed out and breed out life forms deemed superfluous or threatening to colonial settlement. This phase of ecological imperialism (Crosby 1989) had lasting implications for ecological relations across the entire region. Over a relatively short period of time, the land and natural resource management techniques and hunting practices of settlers decimated central Kenya's wildlife populations and radically altered existing ecologies that had been produced by and for pastoralism in centuries prior.

Chapter 2, "Rewilding," focuses on the independence era, when many settlers in Laikipia began to experiment with rewilding the vast landholdings they previously worked so hard to dewild. During this time, agricultural policies in Kenya shifted away from favouring settlers and beef markets began to deteriorate. A nationwide ban on hunting also came into effect in 1977, adding another challenge to the collective identity of settlers and how they performed their notions of belonging and cultural superiority. In response to these and other events during the independence era, settlers began to transform their working farms and ranches into conservancies that were not only suitable for wildlife but also had the potential to generate revenue through conservation activities and wildlife tourism. In this chapter, we outline how settlers went about rewilding their properties with support from international conservation organizations by restoring habitat and reintroducing species they had all but eradicated from the plateau, and by investing in new security arrangements and technologies. We show that rewilding was not about returning to an ecological state that existed prior to colonial settlement; rather, the ecological aesthetics and relations produced through rewilding were meant to reflect colonial imaginaries of nature in Africa and resolve the crises faced by settlers during this time. Through rewilding, structures of settler colonialism were further embedded into the ecologies of Laikipia.

Chapter 3, "Repeopling," recounts how some pastoralists and aspects of pastoral ecologies began to be incorporated into settler ecologies during the 1990s and early 2000s in an attempt to resolve and make productive long-standing tensions between pastoral and settler-colonial land use and tenure systems. Chapter 3 begins with first-hand accounts of pastoral incursions into settler ranches and conservancies in 2015 and 2017. We use these vignettes to highlight deep-seated and recurrent conflicts that have motivated settler ecologists to bring pastoralists, with their livestock and ecologies, back into depeopled settler-dominated landscapes on certain terms and conditions. Specifically, we highlight two dominant approaches to repeopling that have emerged. First, settler ecologists have incorporated pastoralists and their livestock in small numbers into settler ecologies – for example, by allowing indigenous cattle to graze inside private ranches and conservancies under strict rules. Second, settler ecologists have attempted to expand settler ecologies into pastoralist-dominated community conservancies. We argue that repeopling conservation landscapes fixes pastoralists to the landscape in ways that are instrumental and useful to settler ecologies. This chapter reveals how inclusion in settler ecologies can be just as violent as exclusion.

Chapter 4, "Rescuing," draws attention to the creation of sanctuaries for animals that have been abandoned, injured, or belong to highly endangered species. Sanctuaries have recently become a driving force behind conservation action in Kenya and a significant source of finance and revenue. In these spaces of refuge, pretences of wilderness are eschewed to create opportunities for intimate encounters between tourists, researchers, and rescued animals. In this chapter, we situate the conservation-by-sanctuary movement in relation to historical acts of rescuing by settlers, which tended to be carried out by settler women who adopted animals often orphaned by the guns of men in their families. The knowledge and tools developed by these women over the years informed sanctuary models that are now quite masculinized – and often commodified and militarized. We suggest that at least three modes of rescuing are now predominant in Laikipia: rescuing to release, rescuing to relocate, and rescuing to retain. Each of these work in different ways to nurture quasi-wild animals that – whether released, relocated, or retained in place for breeding and research – necessitate ongoing interventions by settlers and legitimize their authority over wildlife management and conservation. In addition to providing examples of how individual animals can be enrolled in and subjected to settler colonialism, this chapter is unique in critically engaging with the implications of sanctuaries, which tend to be overlooked compared to spaces like national parks, conservancies, and zoos.

Chapter 5, "Scaling," considers the most recent ongoing phase of ecological transformation in Laikipia. With the threat of a mass-extinction event looming, radical plans are being proposed and implemented globally to secure as much of the earth's surface as possible for biodiversity. These include propositions that would see 30 to 50 per cent of the earth's surface set aside for conservation. In line with these trends, Kenya has made the creation of wildlife corridors and dispersal areas in human-dominated landscapes a pillar of its national development strategy. In this chapter, we draw attention to three scaling initiatives underway in Laikipia and adjacent areas that are meant to enable the movement and migration of wildlife throughout the wider landscape: the expansion of the community conservancy network; the establishment of new wildlife corridors; and the creation of an ecological network for rhinos, called Ukanda wa Vifaru. We unpack the expanding and increasingly diverse assemblage of settler ecologists involved in conservation connectivity, drawing attention to national environmental and wildlife authorities. As we show, because settlers already control large, interconnected wildlife-friendly properties in Laikipia and have ready-made access to international networks and novel sources of finance, they are playing important roles in realizing landscape-level initiatives. Finally, we bring conservation connectivity into dialogue with other approaches to ecological connectivity at risk of being erased and replaced as a result of the scaling of settler ecologies.

To conclude *Settler Ecologies*, we respond to a series of questions that highlight how the book as a whole contributes answers. We ask, "Who are settler ecologists?" and offer detailed insights into who, exactly, falls into the expanding assemblage of actors we describe as settler ecologists. One of our key messages is that not all settler ecologists are white Kenyans, nor are settler ecologists always conscious of how their efforts contribute to reproducing structures of settler colonialism. We also speak to the question, "Where are settler ecologies located?" and identify other places where similar phases of ecological transformation might also be observed. In response to this question, we also speak to the uniqueness of Kenya as a case study and the contributions it stands to make to broader knowledge. We then ask, "Should we study settler ecologies?" and consider the risks and opportunities associated with making settler ecologies the focus of analyses in this and other contexts. Finally, we ask, "What are ecologies otherwise?" and discuss why settler ecologies should not be seen as the beginning or end of ecological relations and transformation in settler-colonial contexts.

1 Eliminating

It is April 2023 and we are making our way northwest across Laikipia, travelling a murram public road that cuts through Ol Jogi – a large, private conservancy. We are on our way to Waso, located in the arid and semi-arid rangelands of Naibunga to the north of Ol Jogi. This part of Laikipia has been hit hard by successive years of drought. Although the rains have finally arrived elsewhere on the plateau, any evidence of recent rain here is scant. As we drive, large plumes of dust are propelled high into the sky by our rumbling tires. There is almost no grass to be seen and even common, hardy plants, such as buffalo thorn (*Ziziphus mucronata*) and whistling thorn (*Vachellia drepanolobium*), are atypically sparse. One plant, however, seems to be doing quite well despite the conditions. Much of the earth is covered by a distinct succulent known as *Opuntia* or prickly pear. The flat, oval stems of this cactus – loaded with long, needle-like spines – can be seen for miles in some areas, reaching up towards the sky out of the dusty soil (see both photo 1.1 and the image on the cover of this book).

Although difficult to verify with absolute certainty, it is widely agreed that various species of *Opuntia* were imported to Kenya from South America by white settlers in the 1950s. Some say the plant was first brought to a colonial homestead in Dol Dol, about 12 km from where we are now, as an ornamental potted plant that could survive in arid environments; other reports state that colonial administrators used the cactus to construct living fences around their office buildings (Strum, Stirling, and Mutunga 2015). Some of the *Opuntia* species introduced to the region have since died off, but *Opuntia stricta* has thrived. This species requires very little water to grow, which has allowed it to spread with ease across Laikipia. In more arid regions like Naibunga, the cactus has come to dominate rangelands. In fact, in 2018, the US Forest Service was contracted to carry out a study on the spread of *Opuntia*

Photo 1.1. *Opuntia* in Naibunga Community Conservancy, Laikipia

in Naibunga Community Conservancy, which revealed that the species is present on 90 per cent of the conservancy's landscape and completely dominates vegetation on 12 per cent of the land (Butz et al. 2018).

Opuntia requires very little water to take root and spreads rapidly, displacing and suppressing other species – including grasses, which are important for livestock and wildlife. In the absence of other options, especially during droughts, prickly pear fruits are desirable to grazing animals. The reddish-purple fruit seems to have minimal impact on elephants, baboons, and birds, which all propagate the seeds after feeding. However, the fruit poses grave injury to cattle, goats, and sheep. The cactus spines lacerate the mouths of foraging livestock and can become lodged in their eyes, leading to blindness. Livestock also have trouble digesting the seeds of the fruit, which may clog the intestines of goats and sheep, leaving them unable to feed and ultimately causing their demise. In places like Naibunga, where livestock keeping is people's main source of livelihood, this ecological legacy of colonization has become a persistent thorn in the flesh of pastoralists and is of growing concern as the plant continues to reproduce and spread. Many across the conservancy are attempting to take action, removing the plant and replacing it with indigenous grasses and shrubs, but *Opuntia* remains difficult to control.

As we drive on towards Waso, we pass through areas where local residents and civil society organizations have been carrying out this work. We notice that potholes in the road have been filled using prickly pear plants removed from the surrounding rangelands. This is one way *Opuntia* is being put to practical use after it is uprooted. However, it is also a symbolic testament to the durability of prickly pears as a plant, species, and colonial legacy – and of the desire some have to see the presence of *Opuntia* stamped out of the landscape.

Much of the existing work on settler colonialism in Africa draws attention to the durability of settler-colonial structures through political institutions, land policies, and the law (Bhandar 2016; Løvschal and Gravesen 2021; Mamdani 2006). These forces and processes are all undeniably essential to sustaining settler colonialism. Yet settler colonialism is also memorialized and lives on ecologically through more-than-human relations. As the story of *Opuntia* so aptly illustrates, the ecological relations produced through settler colonialism continue to violently suppress, remove, and erase indigenous lives – human and nonhuman – well after formal independence, just as other structural forces do (see also Todd 2022).

In this chapter, we historicise settler colonialism's logic of elimination and endurance across time by analysing processes of ecological transformation that first began to unfold in Kenya with the advent of colonisation and arrival of settlers. Specifically, we draw attention to three ways that white settlers set out to eliminate species and ecological relations predominant in Laikipia that supported or were sustained by Indigenous pastoralists: settlers cleared and culled occupied land of undesired animals and plants; they hunted wildlife for sport and to earn a living; and they implanted and imported desired species to weed out and breed out less desirable life forms and those that posed a threat to settler economies. Each of these sets of practices shows how the eradication of indigeneity that is fundamental to settler colonialism (Wolfe 1999, 2006) extended to the ecological realm. Our analysis in this chapter builds on Crosby's (1989) insights into "ecological imperialism" by demonstrating that expressions of ecological imperialism did not end with the British Empire. Instead, we argue that the elimination of existing ecological relations from the landscape provided the building blocks for the creation of new settler ecologies that have profound and enduring effects on ecological relations in Laikipia today.

We begin this chapter by recounting key historical events, policies, and processes that led to the dispossession of Indigenous Peoples from Laikipia and creation of the White Highlands. As this has been discussed at length by others (see Hughes 2006; Kantai 2007; Kanyinga 2009; Letai 2015; McIntosh 2017; Fox 2018b; 2018b), we keep this discussion brief. Instead, we focus our discussion on the ecological relations that settlers began to unmake and remake after their arrival to embed settler colonialism in the landscape. Our analysis highlights the ways that clearing, culling, hunting, and implanting were pursued to remove and erase indigenous species. We also indicate how these everyday practices and processes served broader settler-colonial interests and laid the ecological foundations for a durable and long-lasting settler society.

The Creation of the White Highlands

A common sight in the houses of white Kenyans is a collection of books that recount and romanticize colonial relations in Kenya – possibly stacked on a teak coffee table or arranged on a bookshelf of dhow wood, surrounded by other signs of African material culture stripped of meaning. The white settler canon in Laikipia includes a few mainstays, such as *The Flame Trees of Thika* by Elspeth Huxley and *Out of Africa* by Karen Blixen. Many will also have at least one book that retells the story of the creation of the White Highlands. This story almost always begins with the (in)famous trek of Scottish explorer Joseph Thomson through Maasai land.

In 1883, Thomson set out from the Kenyan port of Mombasa with 140 porters and KSh 2,000. Sponsored by the Royal Geographical Society, Thomson was tasked with ascertaining "if a practical direct route for European travellers exists through Masai country from any one of the East African ports to Victoria Nyanza, and to examine Mount Kenia [sic]" (Thomson 1887, 6). Thomson's expedition paved the way for settlers to enter Laikipia, with some of the region's earliest settlers literally following in Thomson's footsteps (Huxley 1953; Trzebinski 1986; Morgan 1963; Duder and Youé 1994). In many ways, Thomson's expedition serves as the creation account in Kenya's white folklore. It has come to symbolize the ordained arrival of white man in "god's country" (Huxley 1948, 1953, 1991; Trzebinski 1986).

Thomson's role in the colonization of Kenya extended beyond the physical mapping and exploration of the interior and its highlands. The explorer's representations of Maasai land and culture also provided the colonial administration with the empirical justification it needed to settle Kenya. Based on his expedition, Thomson asserted that the existing population in Laikipia, the Laikipiak Maasais, had been all but wiped out during the Iloikop Wars, rendering vast parts of the plateau empty and idle. Thomson wrote, "the greater part of Lykipia [sic] – and that the richer portion – is quite uninhabited, owing, in a great degree, to the decimation of the Masai of that part" (1887, 238). The writings of Elspeth Huxley, perhaps one of the most influential contributors to Kenya's white folklore, reflect this same erasure of existing populations from the plateau at the time of white settlement. As Huxley (1953, 73–4) writes in *White Man's Country: Lord Delamere and the Making of Kenya*,

> It was inevitable ... that the emptiness of the land should be the first feature to strike and even astonish the European eye. A man could walk for days without catching sight of a single human being save perhaps for a wild little Dorobo hunter ... or a slim Masai herdsman standing alone with his spear and his sheep on a plain that stretched to meet a fair horizon. It was only natural that some of the early-comers should begin to ask for the land.

Colonial narratives and tropes about the emptiness of Laikipia served as a discursive manoeuvre to justify Laikipia as a place awaiting transformation. As previously discussed, for settlers shoring up access to territory was almost always made possible by *terra nullius* – claims that lands to be occupied are empty, unused, and awaiting settlement. In historical writing by Kenya's settlers, the description of Laikipia as a place lying empty was routine.

These early representations of Laikipia as a vast, rich plateau lying idle in the highlands served as justification for transforming Laikipia into a white reserve where land could be put to what the empire viewed as sound economic use (Huxley 1948, 1953; Collett 1987). Unlike colonial writing about many other parts of the African continent at the time – such as writing on the continent's interior, which had become known as a white man's grave (Izuakor 1988) – Laikipia was seen as ideally suited to the livelihoods and well-being of white settlers. As one settler explained to us, standing on the ranch he inherited from his grandfather, "When my grandfather first came here, he wrote home saying this was God's country." Settlers in this part of Kenya have often relied upon such tropes when discussing their attachment to occupied land, describing Laikipia as "God's country," "a white man's country," and a "park like country."

The idea that Laikipia was made for white settlers is reflected in writing from this time period. For example, a 1935 biography of Lord Delamere – who was one of the earliest settlers in Kenya and a British Peer of the Realm who became unofficial leader of the colony's white community – describes Delamere's first impressions of Laikipia upon ascending into the highlands at the end of a hunting safari that took him through British Somaliland into the northern frontier of the Protectorate. The biographer, Huxley (1953, 53–4), writes,

> The open, wind-swept downs of Laikipia, running from the green foothills of Kenya's mighty peak to the cedar-forested slopes of the Aberdare mountains, are lovely enough to an eye attuned to the beauty of an English landscape. What must they have seemed like to an eye which, for over a year, had seen nothing but stretches of vicious thorn, wastes of lava rock, rivers that were waterless and sandy courses, trees that were stunted acacias and fleshy euphorbias? To be able to drink clear, cool water from running streams congenial to the alien trout rather than to the indigenous crocodile ... Here, indeed, he [Lord Delamere] must have thought, was a promised land.

Huxley's portrayal of Delamere drinking from "streams congenial to the alien trout rather than to the indigenous crocodile" naturalizes white presence (the trout) in the landscape while alienating Indigenous presence (the crocodile). Such discursive manoeuvres serve to erase and deny Indigenous presence and claims to land.

By 1895, just twelve years after Thomson's expedition, Laikipia was declared part of the East Africa Protectorate (Izuakor 1988). One of the individuals responsible for turning nascent visions of Laikipia as a white man's country into reality was Sir Charles Eliot, who was appointed Commissioner of Kenya in 1900. Eliot shared Lord Delamere's ambition of proving to the world that East Africa was ideal for a settler society. Delamere "wanted [Kenya] to become a true British settler colony in the sense that Australia and Natal [South Africa] had been colonies – places where people settled for good and tried to build a replica of England" (Huxley 1953, 95). Eliot, too, saw Kenya as "admirably suited for a white man's country" and was responsible for initiating the policy of settling whites in the highlands (Eliot, as cited in, Izuakor 1988, 318).

If Kenya was to make the jump to a settler colony at this point in time, land had to be readily available to prospective settlers. The Crown Lands Ordinance (CLO) was introduced in 1902 and gave the Crown the power to alienate land, drawing all unoccupied land under the purview of the colonial administration. Ultimately, the Crown would not recognize any claims to so-called unoccupied lands made by Africans (Hughes 2006). The CLO, derived from Canadian land laws, allowed the colonial administration to provide freehold titles – which could be used as collateral for bank loans –for ninety-nine-year leases to white settlers, who could also purchase freehold plots of 1,000 acres or less (Huxley 1953). One of the CLO's provisions was the Canadian homestead principle. This principle allowed successful land applicants to acquire 160 acres initially, becoming eligible for another 160 acres should they successfully meet the administration's development criteria (Huxley 1953). "This process could be repeated twice more, until the grand total of 640 acres, or one square mile, was reached" (Huxley 1953, 85). Following the introduction of the CLO, Kenya's white population increased by 520 people by 1903, and 1,554,000 ha of land was alienated under the CLO before 1915 (Nicholls 2005; Hughes 2006).

Eliot's efforts to recruit settlers to Kenya and boost the Protectorate's economy proved to be relatively successful. Following his departure in 1904, the Foreign Office – replaced by the Colonial Office in 1905 – pursued the now-infamous Maasai Agreement of 1904, in which Maasais allegedly agreed to be moved into native reserves. Eliot had dissuaded such forms of segregation during his tenure as commissioner, as he believed that assimilating Maasais into European society would force Maasais to "become civilized" (Nicholls 2005; also see Hughes 2006). However, his successors had a different approach in mind, and set out to push Maasais off their land and into designated native reserves. These efforts involved two relocations over the course of a decade that were anything but straightforward.

The Maasai Agreement of 1904 was initiated by Sir Donald Steward, Eliot's successor, and was signed by Maasai leader Olonana and other

Maasai representatives. By signing the agreement, Olonana and the other signatories agreed that any Maasais occupying land in the Rift Valley would move into two reserves – one to the north, in Laikipia, and another to the south, along the border with German East Africa (now Tanzania) (Hughes 2006). These reserves were promised in perpetuity to Maasais and compensation was guaranteed should any future changes occur. Little to no resistance was documented, but most Maasais were opposed to the move (Morgan 1963; Hughes 2006). Hughes (2006) discusses evidence that Olonana had been on salary with the British since 1901 and had signed on to both the treaty and moves before his fellow Maasai representatives agreed. Hughes (2006) also suggests that other Maasai leaders were, in some cases quite literally, forced to put their thumbprints on the agreement.

The Maasais that were moved to the northern reserve in Laikipia used the plateau's rich pastures to their advantage, tripling their live-stock numbers in just five years (Hughes 2006). However, this situation would not last long. In 1910, the colonial administration defied the 1904 agreement and initiated a second move, one that sent Maasais that had been moved to the northern reserve in Laikipia to join those that had been moved to the southern reserve. According to the District Commissioner for Laikipia at the time, the second move was designed to "free Laikipia from Masai" and to open additional parts of the plateau for white settlement (as cited in Hughes 2006, 38). However, this reason was not publicly admitted. Rather, the move was said to meet the interests of Maasais and the wishes of Olonana (Hughes 2006).

With these moves complete, Laikipia was "free" of Maasais by 1913 for settlers that had "cast envious eyes at the grazing grounds of Laikipia" (Huxley 1953, 265; see also Duder and Simpson 1997). However, even though land was available, the pace of settlement in Laikipia lagged. In 1915, the colonial administration once again updated the CLO hoping to make Kenya more attractive to would-be settlers. These updates extended the lease period on freehold titles for white settlers to 999 years – from what was previously 99 years – and placed more explicit restrictions on ownership. Only white settlers could own free-hold property, whereas Africans were confined to areas of land within native reserves (Morgan 1963; Izuakor 1988).

By the 1920s, white settlement in Laikipia was starting to gain momentum, buoyed by the Ex-soldier Settlement Scheme of 1919 and further updates to the CLO in 1921. The Soldier Settlement Scheme was particularly transformative, allocating "over a thousand new farms across two million acres of land to British subjects who had given military service during World War One" (Jackson 2011, 347). Although Kenya's highland areas had long been described as white at this point, it was not until the increase in European immigration to Kenya, followed by the Devonshire White Paper of 1923 and the Kenya Land Commission (or Carter Commission) in 1933, that a white reserve was formally established (Huxley 1948; Lipscomb 1955; Morgan 1963). These proclamations from the British government formally excluded Africans from owning land within the boundaries of the White Highlands, formalizing a racialized property regime (Huxley 1948; Lipscomb 1955). Importantly, the report of the Kenya Land Commission also inhibited the formal expansion of white-only land. Thus, in the span of about forty years, a white man's country had been carved out of the highlands, but it was also restricted in spatial terms. This served to make the heart of the colony all the more important to Kenya's white settlers.

In 1905, Sir Charles Eliot said, "The main object of our policy and legislation should be to found a white colony" (Eliot 1905, 103). By the onset of the Second World War, this objective was well underway, as the colonial administration had transformed Laikipia from a highland area that was a territory of Maasais into a demarcated space reserved for white settlement. Many of the discourses, policies, and tactics involved in eliminating Indigenous Peoples and polities from the land base during this time mirrored those used in other settler colonies around the world. Colonial administrators and land laws were often imported from these other contexts, such as Canada, to expedite the expropriation of territory for settler occupation.

The Great Ecological Unmaking

Merely occupying land was not enough to establish a settler colony. This land also had to be transformed from "wasted" space into fertile lands of "useful account" (Huxley 1953, 78). This transformation involved unmaking existing landscapes in the highlands over the better part of half a century. In many ways, this era of colonial settlement was defined by the elimination of existing ecologies and their replacement with those of value to settlers, the colony, and the wider Empire. As the following sections show, eliminating and replacing existing ecological relations featured just as prominently in the creation of the White Highlands as the violent dispossession of pastoralists populations and tales of empty, dead spaces.

Eliminating by Clearing and Culling

As settlers arrived in Laikipia, they were assigned land that may have appeared vacant but played an important role in larger systems of shifting agriculture for Maasais and other groups, who relied on these areas for cultivation and grazing during particular seasons.[1] Throughout the settlement era, it was not uncommon for cultivators or pastoralists to return to these areas after prolonged absences, only to find they had come under the possession of white foreigners (Morgan 1963). Given that fallowing was a relatively common practice in Europe, it seems difficult to believe that settlers could genuinely confuse fallowed pasture for *terra nullius*. Rather, this was likely far more like an intentional discursive manoeuvre to erase Indigenous presence in support of white claims to land.

Regardless, upon acquiring land, most settlers set out to convert dry bushland and shrubland into farms and ranches that resembled the English countryside. Some highland areas were fairly suitable for European agricultural activity upon settlement – particularly those used for agropastoralism in the immediate years prior. However, by and large, before land could be made productive for the settler economy, it had to be cleared. Large swaths of the highlands were cleared of dense vegetation before commercial agriculture could begin, leading to deforestation. A pamphlet prepared for women in new settler households by the East Africa Women's League (EAWL 1953, 46) describes the successes that early settlers had in clearing the highlands:

> Two years ago, this land at Mau Narok looked much the same as the Masai reserve on the other side of the fence – tall weeds choking out the grass with thick clumps of indigenous bush here and there, for the Masai do not normally live at such heights. Today on the European farms rolling fields of wheat stretch to the horizon and pastures support fine herds of cattle and flocks of sheep of European breeds. (EAWL 1953, 46)

Compounding the pressures on shrublands and forests created by colonial settlement and expansion, settlers were also encouraged to freely use trees to construct furniture and houses, and companies were established by settlers to supply timber. Within decades of their arrival, settlers had cleared hundreds of thousands of acres of forested land for agricultural and industrial development (Ofcansky 1984).

---

1 Here, it is important to recognize that agropastoralism and hunting and gathering by these and many other groups with presence in the highlands constitute aspects of the socioecological systems that we oversimplify as "pastoralist."

It was not only vegetation that was cleared by settlers. Animals, too, were removed from the landscape. In the same pamphlet prepared by the East Africa Women's League, entitled *Are You Coming to Kenya? A Guide for the Woman Settler*, the contentious relationship between early settlers and wildlife is highlighted:

> Developing a farm here means taming part of Africa and Africa is not easily tamed. In many of the roughly-built farm homesteads leopard skins on the floor tell their own story. Families of colobus monkeys chatter in the gloom of the bush edging the wheat fields and the farmers must keep a constant eye open for the stray ostriches and zebras who can damage a whole field if allowed to roam unchecked. (EAWL 1953, 46)

As Huxley writes about the early days of settlement, "It became obvious that if cereals were to be grown at all the game would have to go" (Huxley 1953, 161). Beside raiding crops, settlers also worried that wild animals would spread disease and prey upon livestock (Matheka 2005). Stories of violent attacks on settlers by wild animals were also common, including accounts of settlers being mauled by lions or narrowly escaping unexpected encounters with buffalos and rhinos (Duder and Youé 1994).

For these reasons, settlers extensively culled wildlife. Early settlers had no qualms about hunting and shooting wildlife, seeing this as a necessary exercise in the conversion of unruly shrublands and bushlands into productive farms and ranches. Lord Delamere "set the tone in calling for and executing the elimination of wildlife from his estates" (Steinhart 1989, 253). Despite being known as a "gentleman hunter," Delamere also cleared game off his ranch with a Maxim gun (Steinhart 2006). Other settlers followed suit, taking great enjoyment from ridding their farms and ranches of pests and vermin. Some settlers we spoke with in Laikipia expressed bemusement about their ancestors' enthusiasm for shooting wildlife. A settler in his sixties described his father's passion for killing black rhinos, which were once so numerous they were considered pests: "rhinos were like fucking rabbits," he exclaimed.

At times, the killing of wildlife was also pursued out of vengeance rather than self-defence or sport. Acts of violent retribution directed at wildlife occupied a significant place in the settler moral code. In a memoir by Huxley (1959), an English settler named Tilly – a character based on Huxley's real-life mother – describes leopards as "one of the country's natural hazards." In the story, a child, based on Elspeth Huxley, suspects a resident leopard of eating her pet duiker, "Twinkle," as well as being involved in the more serious crime of eating off domestic goats. She describes, at length, a hunt that was organized to shoot and kill the suspect. The settler-hunters involved were not just interested in killing the leopard to protect goats; they also approached the hunt as an act of retribution for its wrongdoings.

Acts of retribution against certain animals is something that continues today – at a smaller scale and in a more covert fashion – on settler ranches. While visiting one ranch in Laikipia in 2015, we listened to a tale about a pet dog that recently died after a puff adder bite received on a morning walk with the ranch owner. Over subsequent days, the ranch owner enlisted the help of friends to track and kill the snake suspected of biting the dog. Whether the actual culprit or not, one "puffie" eventually paid the ultimate price in retaliation. We were even shown proof by the storyteller: a photograph of the dead reptile hanging from a stick. During a later stay at the same ranch, we were charged by a buffalo who entered our camp at dusk. When informed about the incident, the ranch owner explained the young bull had been causing problems around the camp in recent months and, if the buffalo refused to move on, they would request permission from Kenya Wildlife Service (KWS) to shoot the animal. Although there is a national hunting ban in place, settlers with wildlife operations on their properties (unlike Indigenous pastoralists) are still granted special permission to kill certain animals – for example, if a wild animal becomes a problem or has been mortally wounded. These stories reflect how the logic of elimination embedded in settler colonialism remains entwined with everyday attitudes, behaviours, and more-than-human entanglements within settler ecologies today.

During the colonial era, settlers were more or less able to shoot-at-will and shoot-to-kill, as wildlife threatened the economic development of the highlands and the so-called civilized society that settlers had been brought to develop (Collett 1987; Bull 1988; Steinhart 1989; Matheka 2008; Kabiri 2010). Virtually "any animal found on private (i.e. settler) land was fair game that could be killed with impunity and without a game license by the landowner or his agent, while most settlers also possessed a license which allowed the killing of animals on Crown Land" (Steinhart 1989, 253). Additionally, the colonial Game Department provided game control as a service to settlers. "In these control operations," Steinhart writes, "hundreds of animals, especially elephant and rhino, were killed each year to prevent (or to avenge) the destruction of crops or fences on settler farms" (1989, 253). Similar government-sponsored campaigns took place to control zebras and drive other types of game off settler land (Duder and Youé 1994, 266). After all, one of the primary motivations for establishing the White Highlands in the first place was to ensure that fertile parts of the colony served the interests of Empire. If settlers were to take up land in Laikipia and make it productive, they had to be free to defend their property, crops, and livestock from wild animals without worrying about cumbersome legal processes or reprisal.

Although managed very differently, settlers also culled native livestock in the highlands. Settlers looked down on native livestock, arguing that it was poor quality and produced low quantities of milk (Kenya Land Commission 1933). Many also feared that native livestock carried diseases that would spread to their herds and interfere with their plans to export meat internationally, which required strict quality control (Watson 2014; Waller 2004). In response, settlers took multiple steps to limit numbers of indigenous livestock in the highlands. Indigenous herds were moved off settled land – or land desired for settlement – through the Maasai Agreements of 1904 and 1910, with 175,000 cattle and over one million goats and sheep recorded leaving Laikipia in the years following the 1910 agreement (Hughes 2006). The alienation of migratory routes and seasonal grazing areas and pastures to enable European settlement also had protracted effects on indigenous livestock herds, undermining systems of transhumance pastoralism. Eventually, the colonial administration would promote a closed frontier program that prevented indigenous livestock from entering the White Highlands altogether (Overton 1987). Fences were constructed, quarantine stations established, and colonial officers tasked with keeping pastoralists and their livestock away from settler properties and herds. These new restrictions on pastoralists served as additional barriers and strengthened existing boundaries that further divided and segregated the highlands along racial lines. They also extended segregation to the more-than-human realm, establishing clear lines of separation between preferred breeds of livestock held by whites and those maintained by Africans.

Yet, early efforts to control the herds that belonged to Indigenous pastoralists ultimately failed to curb the growth of livestock populations outside settler farms and ranches, and settlers grew increasingly paranoid about "trespassing stock from the overcrowded African reserves" (Anderson and Throup 1985, 331). Throughout the 1920s and 1930s, fears of overstocking blended with heightened anxieties about soil exhaustion and erosion. Indigenous populations shouldered the blame for what was commonly described as an evil occurring in reserves (see DOA 1930; Champion 1933), while little attention was paid to deleterious impacts of intensifying settler farming and ranching practices. For example, a paper published by the Royal Geographical Journal reflects a common sentiment at the time:

> The agricultural methods of the natives are very primitive, and from time immemorial have proved destructive to the areas which they occupy ... It is therefore little exaggeration to say that up to fairly recent times almost every tribe has periodically moved forward, leaving behind it an area the productivity of which has been ruined by their destructive methods. (Hobley 1933, 140)

The writings of Elspeth Huxley reflect this consensus among settlers at the time. In 1937, she wrote, "What then is to be done? There is only one answer: compulsory destocking, accompanied where necessary by reconditioning of land" (1937, 368).

Official efforts by the colonial administration to destock and cull African livestock began in the 1930s and continued into the 1950s (Matheka 2005). Targets for livestock reduction were often based on calculations about the carrying capacity of grazing areas within different districts and reserves (Anderson 1984; Anderson and Throup 1985), as well as demand for beef products across the Empire. For example, during the Second World War, surging demand for tinned beef to feed Allied troops gave the Kenya Supply Board "sweeping powers" to acquire livestock (Watson 2014). The implications of these practices were grave in the eyes of many pastoralists, especially Maasais who had already been forcibly displaced to accommodate settler ranching in Laikipia. The seizure and killing of indigenous livestock through compulsory purchase and culling was often seen as overly punitive by pastoralists and led to instances of conflict and resistance (Watson 2014). These efforts to restrict and reduce indigenous livestock represent an extension of the early clearing of vegetation and culling of game carried out by settlers as they sought to unmake existing ecologies and remake ecologies more aligned with their interests.

Eliminating by Hunting

Clearing and culling were not the only way that existing nonhuman species were eliminated from Laikipia. Game hunting by settlers and tourists also played a fundamental role in rewriting the ecology of the region. Unlike the British Colony of Southern Rhodesia, Kenya tended to attract settlers from the upper echelons of European society up until the Soldier Settlement Schemes of the post–World War era (Bull 1988; Steinhart 2006; Jackson 2011). According to Steinhart, "the dominant figures of the Kenya settler community from before the Great War until the post–Second World War era were extremely wealthy, landed, and often titled aristocrats and gentlemen" (2006, 92). Initially, hunting for these settlers was far more than a pragmatic activity, carried out in defense of their livelihoods, properties, or safety; it was sport. Hunting for sport – for its own sake – was one way that settlers self-identified as elite and expressed their notions of superiority over other classes, races, and species in the colony (Steinhart 2006). For many settlers, tracking and shooting game in Kenya's highlands was "their chief joy in life" and was pursued with the devotion of a life calling (Steinhart 1989, 251).

Over time, hunting evolved from being primarily an act of leisure and sport to an activity also used by settlers to supplement their incomes. This extra income was needed given how many early settlers struggled to make ends meet due to their lack of training and experience in agriculture and animal husbandry. Many settlers dipped into an economy organized around the trade of wildlife parts, such as the skins of lions and zebras or ivory from elephants and rhinos. Afrikaners in the Nanyuki area were seen as making a particularly valuable contribution to the settler economy by poaching and trading game and animal parts (Duder and Youé 1994, 266). However, other groups of settlers also participated in the ivory and skins trade. As Steinhart writes, "It was the exceptional district officer or police officer who did not both shoot for the pot and take out an elephant licence to supplement his salary by the sale of ivory" (2006, 252). Kenya's growing hunting economy would serve settlers well, as they struggled from lack of know-how with agriculture but were drawn to the romance and adventure that they associated with hunting, tracking, skinning, and bushcraft (Steinhart 1989).

Following the widely publicized safaris of President Roosevelt and the Duke of Connaught in 1909 and 1910, respectively, other opportunities emerged for settlers to profit off unfettered access to wildlife and rights to hunt. With demand for hosted hunting safaris on the rise, some settlers shifted away from agriculture and morphed into part- or full-time professional hunting guides (Steinhart 2006). "Some government officers even blamed the safari business for the labour shortage [experienced mainly on settler farms], as hunters often paid double the usual native wage and drew off the most willing workers" (Bull 1988, 197). International clients paid for not just the hunting experience, but also the experience of life as settlers in Kenya – albeit highly romanticized and much more extravagant and luxurious versions than reality. New safari companies emerged to cater to these clients, such as Cransworth's Newland & Tarlton, which was the largest employer in the colony by 1914, booking as many as three hundred clients per year (Bull 1988). These companies provided all-inclusive packages, organizing cross-continental and in-country transport, providing shooting equipment, interpreting for and guiding clients, and arranging the preservation of trophies.

Of course, in addition to bringing more clients and revenue into the country, the number of wildlife being hunted also increased substantially during this time. The role of the settler hunting guide was to ensure that hunting clients were able to shoot the type and quantity of animals that they wished to shoot during their visit. As a pamphlet by Cottar Safari Services from 1937 describes,

> We shoot so many lions that we get tired of telling about them – as everywhere in all East Africa one is liable to run into lions – and get them if in proper hands – Cottar Service has procured more good lions for clients than any other service and with more appeal to the shoots by visiting baits in early morning; over "kills" in the night, raising them by driving the bush-veldt in a car in day-time; with trail dogs, in fact by every known method we get them. (Cottar Safari Services 1937, S/6886)

Many companies would provide guarantees to clients around the types and numbers of animals to be shot during their safari. For example, this same pamphlet by Cottar Safari Services goes on to outline various packages available to clients, beginning with an entry-level package that guarantees clients will shoot one elephant or one rhino, one buffalo, and five varieties of antelope for the cost of €300 during their trip.

This new class of settler-hunters who emerged from the farms and ranches of Laikipia and other highland areas to join the safari economy found the transition between farm life and safari duty both easy and enjoyable:

> Ivory, lion skins and clients were profitable, and a safari was merely an extension of their normal life on farm or ranch. They would no more be without a gun than without their boots, and it was a natural step from shooting lion to protect one's cattle to hunting lion with well-paying clients. When the coffee failed or the rams died, it was the wild animals that paid the bank loans. (Bull 1988, 197)

Settler-hunters experienced relatively few barriers making the transition into the safari industry if their farming and ranching activities failed. In 1913, a simple £50 licence permitted safari clients to kill or capture two Cape buffalos, elephants, hippos, rhinos, and zebras in addition to almost two hundred antelopes or gazelles. Because many other species – including lions and leopards at times – were classified as vermin, settlers could take safari clients to hunt and shoot animals on their land or in the surrounding areas without fee or penalty (Trzebinski 1985; Bull 1988). The relatively low financial costs of getting into the safari business allowed most settlers to combine their passion for safari with the knowledge and skills they acquired from culling and hunting on their own land (Bull 1988; Steinhart 1989, 2006; Herne 1999; Prettejohn 2012).

Eliminating by Implanting

Settler culling and hunting practices devastated the colony's wild-life populations in a rather quick and spectacular fashion. However, "hunting was merely the sharp, cutting edge of the sword. The heavy mass destructive force was created by the clearing of animals' habitats and their conversion to privately-owned farmlands and grazing areas, fenced and cleared of all but the smallest animals and birds" (Steinhart 1989, 252). As Anderson writes, "the industrious white colonizer" set out to build "a little piece of England in a foreign field" (Anderson 2005, i). Upon acquiring land, settlers worked to transform bushlands and shrublands into the orderly, controlled, and intensive agricultural landscapes they desired. After clearing land and controlling pests, they planted new crops and introduced new species. In this regard, settler hunting went hand-in-glove with other sanctioned changes to the ways land could be accessed and used (and by who) that altered ecologies and ecological relations permanently.

In the earliest days of colonial settlement, low-value cereal mono-culture cropping was the norm. Important early cereal crops for the settler economy included wheat, maize, barley, and oats. In 1904, Lord Delamere began commercial production of wheat and, by 1906, he had planted over 1,200 ha (Makanda and Oehmke 1995). As more land was alienated by the Protectorate, more wheat was also planted. By the 1920s, wheat had become a commercially viable crop in Kenya, covering over 30,000 ha of the highlands (Makanda and Oehmke 1995). Yet profit margins were small, and the continuous cultivation of single crops – a practice that intensified due to demand for certain cereals during the world wars – contributed greatly to soil degradation and erosion across the most fertile parts of the colony (Anderson 1984).

Later, Kenya's agricultural sector would be restructured. Settlers were encouraged to diversify from monocropping into mixed farming and to cultivate high-value crops for export. Higher value crops were slowly introduced from the 1920s onward with production and marketing support from the colonial administration, loans from the United States government, and the extension of the Uganda Railway to Mount Kenya in 1931. For example, several highland settlers began to experiment with crops of *Pyrethrum* (now classified as *Chrysanthemum*) – a flower used to create a plant-based insecticide. By the 1930s, the highlands had become the centre of the African *Pyrethrum* industry, with an estimated 15 tons of flowers being produced each year (West 1959). Coffee, tea, and sisal also became important settler crops – all of which were protected and could only be legally planted by white settlers. By the end of the Second World War, the amount of cultivated land in the highlands had grown exponentially through the establishment of cereal farms as well as new plantation farms growing speciality export crops.

New tree species were also introduced during this time. The Forest Department justified the introduction of new species on the grounds that indigenous species were "bad natural reproducers and slow growers" (Leader of British East Africa, 13 May 1911, as cited by Ofcansky 1984, 138). A variety of exotic trees were brought to Kenya to be experimented with on plantations, including black wattle, Guatemala cypress, *Grevillea robusta*, *Syncarpia glomulifera*, *Tristania conferta*, eucalyptus, and casuarina (Fanstone 2016). Some of these species failed. For example, the department reported no success in growing pines (Fanstone 2016). Other species took off, such as black wattle, which was high in demand globally for its dry bark, which could be used as a tanning agent.

Alongside the introduction of new plants, new types of livestock were also introduced and selectively bred into the landscape as settlers attempted to establish a commercially viable livestock sector and dairy industry on specialized large-scale farms. Early settlers built their herds by trading with pastoralists (Waller 2004). Over time, settlers began to develop the beef and dairy industry using selective breeding, importing sires from abroad and crossing them with indigenous breeds, such as zebu, to improve the quality and yield of meat and milk and to boost immunity to local diseases (Huxley 1953). A range of breeds were imported from across the Empire, including Ayrshires, Friessians, Guernseys, Sahiwals, Shorthorns, and Red Polls from Britain, Australia, and South Africa. As a result of crossbreeding, new breeds emerged, such as Boran cattle – now a preferred breed in East Africa's drylands. Other types of livestock were also introduced on settler farms and ranches, such as sheep for wool and mutton. As with cattle, sires were imported and then selectively bred with local breeds. As early as 1904, settlers paid for shipments of pure-bred Merino ewes from New Zealand and English sheep, including Corriedale, Romney Marsh, Suffolk, and Hampshire Downs (Njuguna 2019). The keeping of imported poultry, such as Rhode Island Reds and Leghorns, and pigs also became common.

Building herds through selective crossbreeding was a long and slow process, as imported breeds did not always do well in their new environment and often succumbed to illness (Waller 2004). Some settlers attempted to address this problem by altering the environment so that it better resembled where they had come from; for example, Delamere imported domesticated grasses from England – including seeds of clovers and lucernes – with the goal of improving his pastures and increasing the survival rate of his herds (Huxley 1953). Over the next few decades, settlers filled the landscape with "big-boned, heavy English animals" in place of "small, skinny native animals" (Huxley 1953, 138). In fact, white settlers in Laikipia would go from owning no livestock to 10 per cent of the livestock in the entire country, with over half of this cattle being "grade animals (native/exotic cross-breeds), and the proportion increas[ing] to just under three-quarters by the Second World War" (Waller 2004, 49).

Beyond livestock, other animals were brought into the landscape. For example, trout were imported from Loch Leven in Scotland in 1905 by boat and ox cart. Trout were introduced in reservoirs on farms and ranches and some settlers began small fisheries with the intention of "filling an empty niche in the sporting fish of this Colony" (Copely 1940, 345). Over the years, trout distribution in the region would expand considerably. Today, streams on Mount Kenya can be found teeming with brown trout descended from fish stocked by colonial settlers and populations of rainbow trout stocked in lakes higher up the mountain are growing. Local fisheries and tourist ventures have capitalized on this, such as Trout Tree Farm in Nanyuki and Ragati Conservancy on Mount Kenya, attracting fishing enthusiasts to the region with photos of 5 lb trout waiting to be caught.

Before concluding, it is worth noting the place of women in implanting new species to aid in the creation of settler ecologies. Both white European and Indigenous women tend to be cast in peripheral roles in discussions about Kenya's colonial settlement and are sometimes absent from white folklore, historical records, and academic sources altogether. Their absence from the history books reflects the enduring privilege of white men – a deep-seated privilege forged at the expense of virtually all other groups in colonial society. A few white women, though, are immortalized in early settler folklore. These women are praised as pioneers who extended the culture of their class and gender into Kenya's wildernesses by accompanying their spouses and raising their children on the plains of Laikipia. Safaris attended by the Duchess of Connaught in 1910 and the Duchess of York in 1924–5 are significant in this regard, as they became inspiration for Kenya's ideal-type settler woman – women who extended the "gentleness" of their class into the so-called "wilds" of Africa (Steinhart 2006). Their roles reflect similar roles played by white women in settler ecologies elsewhere in the social reproduction of whiter, more affluent bodies and lives in spaces historically belonging to and used by racialized or Indigenous People (Parish 2020).

At the same time, many settler women also played more significant roles in transforming and reproducing ecological relations than existing writing might lead one to believe. Although women were likely less involved in the culling and hunting of wildlife than men, they were expected to contribute to reproductive labour that extended to animals and plants – raising and caring for new species implanted in the landscape. For example, Huxley writes about how Lady Delamere's days were "so filled with looking after pigs, poultry, and later, ostriches, that she had little time" for anything else (Huxley 1953, 147). The pamphlet prepared by the East Africa Women's League for new women settlers reiterates this role for women, describing how European women in the highlands spend their time growing all the vegetables and a lovely variety of flowers, ranging from daffodils to chrysanthemums, that one might expect to find in an English garden (EAWL 1953, 5). Some settler women took over their husband's or father's farms after they passed away. For example, Ethel Powys retained 5,000 acres of the Mau Narok Estate and returned to farm its wheat after her father, Powys Cobb, died in 1957. The role of settler women in producing ecological relations to sustain settler colonialism is something we reflect on more in chapter 4, in our discussion of women's labour in rescuing injured and endangered wildlife and advocating on their behalf.

Finally, just as Thomson's expedition into Laikipia would not have been possible without indentured labour, nor would the clearing of land, culling of game, and organization of hunting safaris; the care and reproductive labour of women settlers would not have been possible without the input and labour of domestic servants and farmhands, who were often women from nearby native reserves. As in other African settler colonies, Indigenous women were essential to the maintenance of homesteads, care of settler children, and raising of farm animals – including control of pests occupying domestic spaces, such as mice, rats, monkeys, and baboons. These individuals sometimes make appearances in settler memoirs and recollections. For example, the passing of an African nanny who may have provided generations of care to settler children may be mourned for many years and spoken of fondly over tea, meals, and special occasions. However, as yet another form of violent erasure, recognition of these carers' personhood outside the nuclei of settler families – and of the *other* family members, households, and systems of production these women were also responsible for – is less often acknowledged by settlers. The abusive and exploitative conditions under which they may have conducted their care and reproductive labour are also rarely discussed. Despite the relative absence of these women's stories in colonial writing, settler ecologies were undoubtedly coercively forged by women's free and unfree labour alongside the labour of men.

Making Settler Ecologies: Elimination

Southeast of Nanyuki Town at the base of Mount Kenya sits a colonial-era hotel that, in many ways, serves as a time capsule for the initial phase of elimination carried out by settlers. The road to the hotel is lined with jacaranda trees – a species that was imported en masse to Kenya from somewhere in South America by a colonial town planner who loved the spectacle of purple flowers that these trees bare for eight weeks each year. Upon arrival, large, imposing gates restrict access to the 120 acre hotel grounds. Once permitted through the gates, after agreeing to spend US$50 minimum per person at the hotel's restaurant, visitors are welcomed by perfectly manicured gardens and lawns, encircled by coffee and tea gardens that stretch up to the misty forests of Mount Kenya. The grounds are visited by the usual characters: baboons, colobus monkeys, bush bucks, and an impressive diversity of birdlife (see photo 1.2). But it is the unexpected species that catch guests' attention. Indian peacocks strut around the hotel entrance while llamas, brought from Peru, walk through the hotel's bar. On the day of our arrival, older children, clothed in equestrian apparel, are led around the property on dark bay horses, while younger children tag along behind them, accompanied by a friendly black Labrador. The grounds surrounding the hotel are undeniably beautiful, yet the ecological relations they are composed of are largely imported and artificial.

Inside the hotel, many of the ecological relations that were eliminated before they were replaced are memorialized. Fondly described as an "old-school colonial setting ... reminiscent of times gone by" (Roffers 2016), the Dutch colonial-style hotel is decorated with sepia-toned photos that depict the escapades of Kenya's white settlers and celebrity safari hunters. Its white walls are also adorned with trophies of the animals they hunted back when this was a hunting lodge. The horns of antelopes, impalas, and gazelle are strategically positioned in corridors. Busts and skins cover the floors and walls, including the intimidating head of a male buffalo that looks so lively it causes our baby to turn and run. Perhaps most unnerving of all is the brazen display of multiple enormous tusks of ivory, which stand like alters in the open courtyard (see photo 1.3). The hotel is a living museum preserving artefacts of settler domination, providing modern-day guests with a rare opportunity to travel back to a time when the ecological violence of colonial settlement was glorified.

Photo 1.2. A colobus monkey and peacock share food on the grounds of Mount Kenya Safari Club

Spaces like this hotel are now few and far between. Very few other tourist properties in the area so proudly publicize their role in all but eradicating indigenous species from the landscape – whether it was for sport and leisure or economic imperative. Rather, most settler-owned tourist ventures hide this part of their past away. The names of white hunters immortalized in Kenya's white folklore, such as Dyer, Grant, Hunter, Prettejohn, and Russell, remain well-known in Laikipia, but tales of their hunting glory days are reserved for their own community. Trophies kept from these days are tucked away in corners of homes, kept out of tourists' gazes and replaced with photos and art of living animals instead. As one settler said to us, flipping through his father's old photo albums, "I don't know what to do with all the photos of dad with dead rhinos."

The enduring and devastating impacts of early colonial settlement on wildlife populations in Kenya is well documented in the literature (Akama, Lant, and Burnett 1996). Near Nanyuki Town, it was not uncommon to see lions before the 1900s, while the sight of lions was scarce by 1926 (Duder and Youé 1994, 266). The hunting practices of settlers led to the near extinction of elephants and black rhinos by the 1930s (Duder and Youé 1994). By the time of Kenya's independence in

Photo 1.3. Tusks of ivory from elephants leading into Zebar at Mount Kenya Safari Club

1963, bongos and leopards were also rare in Laikipia, as well as other endemic species such as elands, gerenuks, giraffes, hartebeests, zebras, and oryxes. Of course, certain Indigenous populations hunted extensively for cultural and subsistence reasons as well. However, the trophy hunting carried out by early colonial settlers was more devastating to Kenya's wildlife than any other form of hunting: this includes hunting practiced by ivory hunters in the late 1800s or the game control services later offered to ranchers by Kenya's Game Department – both of which were also notoriously destructive forms of hunting (Bull 1988; Herne 1999; Steinhart 1989, 2006).

Importantly, the elimination of existing ecological relations to make way for colonial settlement was not carried out through hunting alone. Settlers also eliminated existing ecologies through culling and clearing and sought to replace species eliminated with those imported from abroad or selectively bred to suit their interests, needs, and values. Today, some of the most dominant crops in the highlands are those introduced by colonial settlers. For example, Laikipia continues to produce most of the world's *Pyrethrum* (Kamau et al. 2019) while the coffee and tea sectors contribute significantly to the gross domestic product, foreign exchange, and the direct or indirect employment of millions of people (Karuri 2021). Crossbreeds of imported and indigenous livestock, such as Boran cattle, still produce some of the most desired meat and milk in the country and demand the highest market value. The region also produces substantial amounts of wheat and barley and smaller amounts of timber through plantations of *Pinus radiata, Pinus patula, Cuppressus lusitanica,* and *Eucalyptus.*

Acts of clearing, importation, and implantation played a fundamental role in laying the ecological foundations for a durable and lasting settler society. In Laikipia, where about 40 per cent of land is held by some forty settler families today (McIntosh 2017), the dominance of crops, breeds, and commodities introduced at the time of settlement remains visibly tied to economic power and the distribution and control of land today. For example, the 60,000 acre Sosian Ranch and the 40,000 acre Mogwooni Ranch are owned by settlers and run large herds of Boran cattle. Sosian claims to have the largest registered herds of Boran cattle in the world while Mogwooni's herds of sheep and cattle are reputed to be the best in the country. As with many other family landholdings in Laikipia, settlers maintain their connection to – and, indeed, control over – land by keeping high-value species that their ancestors first introduced to remake the ecology and economy of Laikipia on their arrival.

Even in parts of Laikipia where settlers no longer own or control large landholdings, the ecological imprint caused by colonial settlement remains. Recent studies show that this is the case in post-colonies all over the world, with evidence of colonialism detectable in the ecology of post-colonies centuries after the end of empire. As Lenzner et al. write, "the persistent legacy of human activities on biological invasions over centuries [is] reflected in the compositional similarity and homogenization of their floras" (2022, 1732). This ecological imprinting is not an accident, but rather was always part of the mandate of settler colonialism. As Huxley explains,

> It is sometimes said that if Europeans were to withdraw from Africa today the continent and its people would revert to savagery and all traces of our civilisation would be expunged. This is not altogether true. Whatever the fate of our cultural influence, we should at least leave behind indelible traces of our cattle and sheep in the hereditary mechanism of animals which survived us. We should leave plants that have colonised the soil perhaps more permanently than men – wheat and barley, sisal and coffee, oats and tea, potatoes and peas, fruit and wattle trees. These at least would remain as a memorial to Europe's conquest of Africa. (Huxley 1953, 176)

This defence of ecological imperialism reveals the ontological core of settler colonialism and illuminates the significance of ecology to this ontology. A long history of work on settler colonialism in Africa draws attention to the durability and visibility of settler-colonial structures in legal and political institutions (Mamdani 2006) and land policy (Bhandar 2016; Løvschal and Gravesen 2021). This chapter adds to this literature by showing how settler colonialism lives on through ecological relations as well.

2 Rewilding

Around the time of Kenya's independence in 1963, a period of reckoning began for many settlers concerning their contributions to biodiversity loss and species decline throughout the colonial era. This reckoning was often self-imposed and inward facing, as settlers came to terms with how their culling and hunting practices depleted wildlife populations on a large scale and caused the local extinction of some species. A memoir written by Daphne Sheldrick, who is from a well-known settler family that remains influential in Kenya's wildlife sector, captures shifting ideas and imaginaries of wildlife among settlers during the latter half of the twentieth century. Sheldrick writes,

> How lightly my ancestors shot at animals. For us, now living in a different era, conscious of the decimation of wildlife and privileged even to glimpse such creatures in a wild situation, the actions of my forefathers appear shocking and difficult to understand. But at that time the maps of Kenya showed little on their empty faces, and beyond each horizon stretched another and another endless untouched acres, sunlit plains of corn-gold grass, wooded luggas, lush valleys, crystal-clear waters. And everywhere there was wildlife in such spellbinding profusion that it is difficult for those who were never witness to this to even begin to visualize such numbers. At the time, no one ever imagined that any amount of shooting could devastate the stocks of wild game, let alone all but eliminate it. (Sheldrick 2012, 6–7)

This quote offers absolution to earlier settlers portrayed as simply misguided in their approaches to hunting and managing wildlife. At the same time, it reflects a new sentimentality towards wildlife that became apparent in the discourses and imaginaries of second- and third-generation settlers during the post-independence era.

If the first phase of settler-led ecological transformation in Laikipia was about dewilding, the second phase of ecological transformation was about selectively rewilding. In this chapter, we draw attention to how second- and third-generation settlers and other settler ecologists began to experiment with rewilding their landholdings in the years after independence – transforming former ranches into conservancies that would attract wildlife in large numbers and, eventually, have the potential to generate revenue through conservation activities and wildlife tourism. We use the term "rewilding" loosely to describe any initiative undertaken with the aim of restoring a certain aspect of nature (Jørgensen 2015), while recognizing this approach to ecological transformation was often informed by colonial imaginaries of nature rather than ecologies that actually existed in the past. We show how the rewilding of settler ranches would come to redefine relations between settlers and wildlife in the decades to come, with settlers assuming the role of custodians and guardians of wildlife rather than hunters and adversaries and international conservation organizations providing settlers with the resources needed to step into this new role.

Crucially, we argue that the settler imperative to rewild was not simply motivated by a desire to rewrite their historically antagonistic relationships with wildlife. Rather, it was spurred on by a collective identity crisis that settlers experienced as Kenya's independence neared and passed, along with concurrent political, economic, and social crises at the national level that threatened the settler way of life in Laikipia. These events and trends led to multiple crises for settlers in the form of growing tensions over settler claims to land leading up to independence, changes in economic and agricultural policies, and a nationwide ban on hunting, which altered how wildlife could be used to generate revenue. These historical events came together to influence settlers to take up rewilding as their next project, while also creating the conditions for international conservation organizations to enter the landscape and transform it into a conservation laboratory and world-class safari destination. Notably, it was during this phase of ecological transformation in Laikipia that who counts as a settler ecologist began to expand, as national and international conservation organizations, investors, and philanthropists became instrumental to rewilding. Ultimately, the rewilding phase of ecological transformation – set into motion by new collaborations and partnerships between settlers and international conservation organizations – created altogether new ecologies that enabled settler colonialism to persist in Laikipia throughout Kenya's post-independence era.

We begin this chapter by describing the national parks movement that emerged both in Kenya and globally in the early 1900s as a solution to biodiversity decline and species loss across the wider British Empire, as this helped generate international demand and support for rewilding. We then proceed to discuss several events that threatened settler identity and society in the build up to Kenya's independence. We show how settlers and their supporters went about resolving the crises they faced by using opportunities presented by the national parks movement and growing concern for biodiversity loss internationally, rewilding their ranches by reintroducing endangered and locally extinct species, and fortifying their rewilded land so wildlife populations could grow. We conclude the chapter by reflecting on how rewilding fundamentally altered relations between people and wildlife in the area and expanded the range of investors and partners that could be relied upon to secure and reproduce ecologies that served settler-colonial interests.

Kenya and the National Park Solution

Long before settlers found renewed purpose and a sense of solidarity in their collective efforts to protect wild species, other imperial actors had already grown concerned with the preservation of fauna and flora in the colonies. In the 1890s, British prime minister and foreign secretary Lord Salisbury expressed concern about the excessive destruction of wildlife in Kenya in particular (MacKenzie 1998). By this time, it was already widely accepted that settlers and white hunters had decimated populations of megafauna in North America, such as the plains buffalo, and contributed to the devastation of species caught up in the transatlantic fur trade, such as beaver (Beinart and Hughes 2007). Hoping to turn the tides in what was then still the British East Africa Protectorate, Salisbury requested that the commissioner enact stricter regulations around game hunting, such as higher licence fees, and create new game reserves (MacKenzie 1998). The commissioner, however, proved to be less receptive to the reforms than Salisbury hoped.

In response to the resistance Lord Salisbury faced, he searched for an international solution to the problem of species decline that had become characteristic of British colonies. In 1897, Salisbury began to advocate for the adoption of resolutions by the international community – mainly the imperial powers of the time – to preserve fauna and flora across the African continent (MacKenzie 1998). In April 1900, the first International Conference for the Preservation of the Wild Animals, Birds and Fishes of the African Continent was held in London with attendees from Great Britain, France, Germany, Portugal, Italy, Spain, and the Congo Free State. During the conference, the Convention on the Preservation of Wild Animals, Birds, and Fish in Africa was signed, an agreement that sought to prevent the massacre of wildlife and support the conservation of diverse species (MacKenzie 1998). Just as Lord Salisbury had lobbied the Commission of British East Africa the previous decade, the agreement also encouraged signatories to establish new game reserve areas. Although the convention was never ratified, it was successful in that it informed subsequent legislation that emerged in the colonies, including revised game regulations and newly gazetted game reserves (MacKenzie 1998).

Even if the uptake of game reserves was slow in Kenya, during the 1920s and 1930s, the national park solution became an increasingly common approach to preserving wildlife and their habitats in other colonies, thanks in part to international conservationists who promoted national park systems around the world. Following the national park movement's traction, the Second International Conference for the Protection of the Fauna and Flora of Africa was held in London in 1933. At the conference, colonial powers signed the London Convention – or the Convention Relative to the Preservation of Fauna and Flora in the Natural State – which committed governments in Africa to the national park solution (Jepson and Whittaker 2002). After the London Convention, colonial governments in Kenya, Tanganyika, and Uganda undertook studies to inform the establishment of national parks in their territories. Initially, imperial expenditure on conservation remained tight during the war years. However, following the Second World War, East Africa's national park system began to take shape with the gazettement of Nairobi National Park in 1946 on the former Nairobi Commonage Lands. This was followed by the gazettement of Tsavo (East and West) National Park in 1948, Marsabit and Mount Kenya national parks in 1949, and Aberdare National Park in 1950.

As early as 1930, the Society for the Preservation of the Wild Fauna of the Empire had been calling for the creation of national parks in East Africa that cordoned off nature from humans and their activities (Hingston 1931). The premise behind this form of alienation was that the land use and livelihood practices of many African populations were harmful in their effects on carrying capacity, soil quality, and vegetation cover. The creation of national parks became a means for evicting people from their homes and farms, forests and pastures, mountains and rivers, and moving them to native reserves or new settlements where the ecological stakes were lower and where they could be easily governed. Meanwhile, within the boundaries of newly minted national parks and reserves, nature would be free to go wild – albeit, somewhat paradoxically, under the watchful eye and constant intervention of authorities.

As discussed in the previous chapter, in early protected areas, "eliminating commercial hunting for ivory and skins gave way to hunting as a subsidy for European advance, leading in turn to the development of ritualized and idealized hunting practice by a colonial elite obsessed with trophies, sportsmanship and other ideals of British boys' education" (Adams 2009, 128). The elite, colonial nature of early protected areas is partially illustrated by a now-famous incident in 1952, when Queen Elizabeth learnt that her father had died and that she had become queen while on safari in Aberdare National Park. Indeed, in addition to enclosing land to protect biodiversity and wild species, it was hoped from the beginning that national parks would attract more tourists to East Africa and inject "economic value to the colonies concerned" (Hingston 1931). The well-publicized safaris of film stars and global leaders, often including members of Britain's royal family, supported this objective, just as those of international celebrities and social media influencers do today.

Over time, the primary orientation of many protected areas in Kenya and other colonies continued to evolve, with less emphasis placed on managing game numbers for hunting and more placed on preserving fauna and flora in situ for ecological reasons and for the viewing pleasure of tourists. This is partly because naturalists and conservationists themselves developed greater concern for the extinction of wild species and the humane treatment of animals. Even in the early 1900s, colonial naturalists in East Africa were observing and reporting on the negative impacts of hunting and industrialization on fauna and flora. A series of high-profile extinctions during the second half of the nineteenth century had amplified such concerns (Jepson and Whittaker 2002).

With the establishment of national parks in Kenya during the 1940s and 1950s, the number of administrators, practitioners, and scientists involved in the operation of these spaces also grew. These authorities often had training in the natural sciences and cognate disciplines, making use of the latest Western scientific knowledge, information, and technology to manage wildlife populations for their intrinsic value rather than purely exchange value. For example, in the 1950s, Kenya carried out its first wildlife censuses, including the first aerial surveys of large mammals occurring in specific protected areas – an activity that has now expanded and takes place at a national level today, with the most recent wildlife census completed in 2021.

The growing concern for the health of ecosystems and the rights and well-being of animals changed the nature of conservation in the latter years of colonial rule in Kenya, "introducing an element of sentimentality" into wildlife and protected areas management (Schauer 2015,

196). This sentimentality was not simply an expression of mourning for something lost or a desire to return to an ecological state that once was; it was an exercise in rebranding the relationship between empire and nature, as well as a new avenue for colonizers to perform their notions of cultural and racial superiority. At least part of the justification for enclosing land and establishing protected areas became the belief that Africa – increasingly synonymous with wilderness and charismatic megafauna in colonial and geographic imaginaries – needed to be saved from the Africans (Nelson 2003), rather than from the colonizer, trophy hunter, and settler. This mandate originated in Europe and was eventually accepted and embraced by settlers in Kenya, as well as colonial administrators as they took up the mantle of conservation.

In these ways, the national parks that emerged in East Africa in the first half of the twentieth century can be understood as a manifestation of the dichotomous relationship between nature and society espoused in Western philosophy. When Garrett Hardin's *Tragedy of the Commons* was published in 1968, this only added more fuel to the growing fire behind protected areas (Hardin 1968). Hardin's theory upheld the presupposition that people's propensity to act in their own self-interest would lead them to overexploit and degrade environmental resources to a state beyond repair. Thus, establishing national parks and reserves did not simply entail the relocation of people from one area to another; it was a continual exercise in enforcing the boundaries of protected areas and protecting the species inside them from external threats. For this reason, national parks and reserves would later be described as fortresses and militarized conservation spaces (Brockington 2002; Duffy 2014). Defending the borders of these fortresses also required the establishment of new laws, policies, and procedures that allowed authorities to enforce bans on human settlement in protected areas and carry out evictions, as well as punish or kill those found to be or suspected of being in violation of permissible activities in protected areas.

As the decades went on and protected areas took root – both physically in the landscape and philosophically among the population – Kenya found itself in a bloody internal war against poaching (Neumann 2004). During the 1980s, KWS adopted a shoot-on-sight and shoot-to-kill policy under the directorship of a white settler named Richard Leakey. Under Leakey's leadership, violence and fear of violence were used to defend protected areas and their wildlife. Shoot-on-sight policies permitted anyone found in protected areas suspected of being a poacher to be shot without question. Leakey became known globally for this approach and for his campaign to ban the export of ivory from Kenya – an agenda displayed in spectacular fashion by setting fire to a giant pyramid of seized elephant tusks twice under his leadership. During the 1980s, Leakey's "intent was to build the best-armed, most lethal wildlife security force in Africa," with a focus on protecting elephant and rhino populations (Hammer 1994, para. 22). In the process of reining in poaching, Leakey also attracted millions of dollars of conservation finance to Kenya during his years in office. Thus, as Leakey became something of an international celebrity in conservation circles, he was also accused of "valuing Kenya's wildlife more than its people" in other humanitarian circles (Hammer 1994).

Conservation outside National Parks

In the midst of all this, many settlers in Kenya initially resisted the sentimentalized vision of conservation that had come to dominate national policy and managed their landholdings as they had earlier in the colonial era. More or less at their discretion, they culled and hunted wildlife on their properties (Sundaresan and Riginos 2010). Attempts by the government to encourage the preservation and protection of wild species on large private landholdings were repeatedly met with disdain and resistance by settlers. As Chongwa (2012) writes, even after independence, there remained a nostalgia for hunting safaris in Kenya among both settlers and the rich and famous in Europe and North America. For many years, settler ranches operated as zones of exception, where settler ideologies of hunting persisted even as the protected area solution was being mainstreamed. However, some settlers in Laikipia did slowly begin to change their tune about hunting, protected areas, and the conservation of endangered species – a change that was driven by several key decisive moments in Kenya's national history.

From Cattle Ranches to Wildlife Conservancies

In the 1920s, demands for land reform and redistribution focused on the plateau. During the so-called "Indian crisis,"[1] many in the South Asian population of Kenya – which outnumbered the European population by two to one – demanded the highlands be opened up for their settlement (Duder 1989). To refute these demands, some settlers

---

1 As with all problematic terms and colonial language in the book, we do not refer to the "Indian crisis" uncritically. This was seen as a crisis for its potential to threaten colonial settlement, but it was a crisis caused by the inequalities and injustices of colonization.

made the racist argument they should be allowed to hold on to their land as they had a responsibility to "civilise the native" and Indian presence in the highlands would be "injurious to African interests on both moral and economic grounds" because subordinate skilled and semi-skilled positions would be filled by Indians rather than Africans (Jackson 2011, 348). Ultimately, the British government sided with the settlers through the White Paper of 1923 and the Kenya Land Commission (or Carter Commission) in 1933, formally establishing the highlands as a white reserve and denying Kenyan Indians land ownership.

Managing to stave off one threat to their interests, settlers would face an even greater political crisis just a few decades later. With colonial settlement, land was not just taken from Maasais, but from Kikuyu as well. Settlers had long relied on Kikuyu to work their land, drawing workers to their farms with the promise of wages. However, throughout the 1940s and 1950s, political tensions between settlers and Kikuyus grew, as the availability of fertile land diminished, leaving many Kikuyu families without enough land or wages to make ends meet while also foreclosing opportunities to enter Kenya's growing agricultural markets (Bates 1987). Initial attempts at peaceful negotiations between settlers and Kikuyus failed. By 1952, a state of emergency was declared, marking the beginning of four years of military operations by the British against a growing anti-colonial movement known as the Land and Freedom Army or Land Freedom Army.[2] The Kenya Human Rights Commission says 90,000 Kenyans were executed, tortured, or maimed by the British while 160,000 were detained in what were effectively concentration camps during this decisive moment in history (KHRC 2012). This anti-colonial movement undoubtedly contributed to Kenya's independence in 1963.

Following independence, settlers feared for their future in Kenya – including the risk of land redistribution, which had been a central demand of the Land and Freedom Army – and many debated leaving the country. Yet Jomo Kenyatta, the first president of Kenya, eased

---

2 Dubbed the Mau Mau Uprising by colonizers – standing for *Mzungu Aende Ulaya, Mwafrika Apate Uhuru* in Kiswahili ("let the foreigner go back abroad, let the African regain independence") – the Land and Freedom Army or Land Freedom Army was the name given to this anti-colonial struggle by those fighting against colonial occupation (Boone, Lukalo, and Joireman 2021).

some of these concerns. One of Kenyatta's first political moves was to silence political opposition, including the revolutionaries of the Land and Freedom Army. By the mid-1960s, "not only had Kenyatta succeeded in building for himself a semi-authoritarian pinnacle of power, but he had also become the favoured grand old custodian and guarantor of the white settlers' economic interests" (Wa-Githumo 1991, 11). On the issue of land, Kenyatta promised to respect land rights, regardless of landowners' race (Gibbs 2014). Although some settlers remained worried that Kenyatta would eventually have to give in to pressure to redistribute land from factions within the African nationalist movement, Kenyatta apparently quelled these fears by personally meeting with white farmers and promising to let them keep their land as long as they agreed to excuse themselves from national politics (Gibbs 2014; Fox 2018a).

There were other signals given to settlers, too, that their presence, and the ecologies they destructively created on settled land, would be welcomed in the newly independent country. For example, the variety of plants adorning Kenya's new coat of arms was intentionally designed to be familiar to settlers (Gibbs 2014). Representing important agricultural products of Kenya typical of European farms – including *Pyrethrum*, coffee, sisal, pineapples, and tea – these plants which were "so intimately linked with the European farming community had the potential to comfort white settlers, reassuring those intent on staying and inducing those as yet undecided to do so" (Gibbs 2014, 510). Similarly, six of the twelve commemorative stamps released in the lead up to independence featured economic sectors where Europeans had strong footholds, including tea, coffee, *Pyrethrum*, fishing, timber, and heavy industry. As Gibbs (2014) writes, such decisions would have signalled to settlers that newly independent Kenya intended to have strong continuities with social and economic norms established by the colonial administration.

Still, some settlers chose to stay and take Kenyan citizenship while others chose to leave. Those that left sold off the land that their families had accumulated over the years and returned to Britain or other countries. Many of the larger landholdings sold off during this time were taken by the government, subdivided, and resold to Kenyan farmers on liberal, long-term credits as part of smallholder settlement schemes. For example, much of the land purchased by the government in the former White Highlands became part of the Million Acre Settlement Scheme, in which approximately 1.2 million acres was transferred to African smallholder farmers (Boone, Lukalo, and Joireman 2021). Some settler ranches were sold to other settlers who stayed behind, allowing families to consolidate even larger properties, which would become vital to the emergence of conservancies in the coming years. A new wave of expatriates from Europe, North America, and southern Africa also arrived in Laikipia during this time. Although this new wave of expatriates were not of the colonial era, many naturalized as Kenyan citizens and integrated into Laikipia's remaining settler community.

The post-independence era proved to be a difficult time period for the white owners of farms and ranches in Laikipia. Changes to agricultural policy meant that crop and livestock prices were no longer fixed to work in their interest. International beef markets contracted as a result of falling demand and tightening export regulations to Europe and the Middle East (Witucki 1976; Sundaresan and Riginos 2010). Furthermore, once the nationwide ban on hunting took effect in 1977, hosted hunting safaris were no longer an option for earning extra income. Only independently wealthy settlers seemed able to manage their properties and sustain their livelihoods (Heath 2000). In short, for many white Kenyans, the future did not appear very promising during this time (Doro 1979).

However, the rapid growth of Kenya's international tourism market provided white Kenyans with an untapped opportunity. Between 1958 and 1968, international tourists visiting Kenya increased six-fold, from 41,000 to 262,000 visitors per annum (Jackson 2011). For international tourists, Kenya provided the same prospects of viewing big game as elsewhere in Africa while also being English-speaking and relatively politically stable. Many tourists also sought out the "revivification of the colonial aesthetic" that Kenya provided, allowing tourists to exist in the "fantasia of the colonial past" many decades after independence (Jackson 2011, 356). Moreover, the Kenyan government was supportive of this growing industry, especially as it brought vital reserves of foreign exchange (Jackson 2011).

Few were better equipped to capitalize on Kenya's growing international tourism market than white landowners in Laikipia, and it is partly for this reason that many began to change their tune about conservation during this time. They held large enough landholdings to support wildlife and, in some cases, multiple tourism facilities targeting different clientele. In fact, following the hunting ban, many farms and ranches were seeing wildlife naturally repopulating their land (Georgiadis 2011; LWF 2012). Many of the older settler families were already experienced in the wildlife tourism sector – albeit with hunting safaris rather than eco-tourism and photographic safaris – and therefore had the national and international networks needed to break into the sector. Even smaller and medium-sized ranches without much wildlife on their land initially were able to enter the eco-tourism market by operating affordable, minimalist camps targeting white Kenyans from Nairobi. Finally, neither of the national parks established in the highlands could offer tourists the savannah-like landscapes that were in such high demand in national parks like Amboseli and Maasai Mara in southern Kenya, and Ngorongoro and Serengeti in northern Tanzania. Large cattle ranches on the plateau were able to fill a market gap in the highlands by offering tourists access to iconic landscapes and megafauna that could not be viewed in the same way in the mountainous Aberdare or Mount Kenya national parks.

Furthermore, by this point, second- and third-generation white Kenyans were beginning to play a greater role in managing their parents' businesses and properties. As implied in the quote at the beginning of this chapter, when it came to environmentalism, these younger generations tended to see themselves as more enlightened and progressive than their parents and grandparents. They had fond memories of carefree hunting and safari holidays as children and of being reacquainted with wildlife on their family land after returning from boarding schools in the UK, South Africa, or elsewhere in Kenya. Similarly, the new wave of expatriates from Europe, North America, and southern Africa that had acquired land in Laikipia also tended to hold different views of wildlife and the importance of preserving nature than earlier generations of arrivants. Thus, as the future of their ranches became more tenuous, these white landowners were very much open to replacing hunting with wildlife tourism as a safety net and livelihood-diversification strategy.

Lewa Downs was among the first to fully embrace the shift towards conservation and wildlife tourism. Purchased by the Douglas family in the 1920s through a colonial settlement scheme, Lewa was managed as a cattle ranch for the next fifty years. In 1977, when Douglas's grandson, Ian Craig, inherited the land, Lewa was still operating as a cattle ranch. However, witnessing the decline of the cattle sector and losing a considerable number of livestock to a severe drought in 1984, Craig sold off most of his remaining livestock and began to experiment with conservation enterprises (Thomas 2000). With the support of Anna Merz, a British-born philanthropist and rhino enthusiast, Craig started by fencing approximately 5,000 acres of Lewa land as a rhino sanctuary, called Ngare Sergoi. By the mid-1990s, Ngare Sergoi's rhino population had grown so large that more space was deemed necessary. Craig decided to convert the whole of Lewa into a private conservancy, now Lewa Wildlife Conservancy. Not only did this prevent bankruptcy and the need to sell off land or other assets after the downfall of the cattle industry, but the shift also eventually brought additional land outside the conservancy under the management of Lewa, as the government granted Lewa trusteeship of the nearby Ngare Ndare Forest Reserve. Craig was able to cover approximately 30 per cent of Lewa's expenses with new tourism revenues by the mid-1990s while remaining expenses were covered through conservation finance, donations, and other investments (Szapary 2000).

On the opposite side of Laikipia, along what is now the county's southernmost border with Nyeri, Solio was another settler ranch that led the way in embracing conservation – although this ranch took a slightly different approach to entering the sector. In 1966, an American mining magnate named Courtland Parfet bought Solio, a cattle ranch at the time. Courtland's wife, Claude, persuaded him to fence off part of the land for wildlife (Patton 2010). Although the Parfets continued to rely on beef and milk production as their main source of income, wildlife flourished on their large, secure, and remote property. Soon, Kenya's Wildlife and Conservation Management Department – which preceded KWS – asked the Parfets to take in several black rhinos in need of protection. By 1980, Solio had become the permanent home of twenty-seven rhinos. By the 2010s, the game reserve had the highest density of rhinos in East Africa and was seen by some as the "heartbeat of rhino conservation in Kenya" (MacEacheran 2013, 2). With growing expertise in rhino husbandry, Parfet dabbled in further opportunities to profit from the endangered creatures – for example, by farming rhinos that could be sold to conservancies across Africa or used to stock national parks. According to Van den Akker, Parfet compared this work "with old French families taking down a painting from the wall when short of funds – when he needed money because his cattle ranch did not bring in enough, he simply sold a rhino" (2016, 118). This mentality succinctly reflects the shift that occurred in how settlers understood and related to both conservation and wildlife during this period.

Quite quickly it became clear that there was some profit to be made through ventures that supported conservation efforts and attracted wildlife tourists to former ranches. With Lewa and Solio spearheading the transition from cattle ranching to wildlife conservation, it was not long before others followed – most especially ranches that, like Lewa and Solio, were used for cattle ranching, rather than the coffee and tea estates in Kericho, Nyandarua, Nyeri, Trans-Nzoia, Uasin Gishu, and other highland areas stretching down to Nairobi and the Rift Valley. In the 1990s, a series of reports by the African Wildlife Foundation (AWF) suggested that returns from non-consumptive wildlife uses were beginning to surpass those of livestock ranching on large properties. In fact, returns for engaging in the wildlife sector were around four times higher than for livestock ranching on comparable sizes of land (Elliott and Mwangi 1997). It was also estimated that a twenty-five-bed safari camp in Laikipia could generate net profits of up to US$400,000 per annum. Although it would be quite some time before new tourism ventures in Laikipia saw this type of revenue, signs that the wildlife sector could attract revenue to private conservancies were all pointing in the right direction, including growing interest from donors and investors as discussed in the following section.

Notably, not all newly established conservancies were set up as money-making endeavours. For the independently wealthy, the establishment of wildlife conservancies on their land or acquisition of land to establish wildlife conservancies was influenced by lifestyle and values rather than a need for economic survival. For instance, Ol Jogi, purchased as a private estate by Alec Nathan Wildenstein, a French-American tycoon, was converted to a wildlife reserve in 1986. The Wildensteins, who also owned an adjoining 31,000 ha ranch, developed the land for wildlife conservation and the enjoyment of their personal visitors. Aside from their family and friends, Ol Jogi was open only to a few select staff, including a researcher, a wildlife veterinarian, and several rangers. It is reported that funds to develop and sustain Ol Jogi came from the Wildensteins themselves until the conservancy changed hands in 2013.

Thus, in Laikipia, growing tensions around claims to land, struggles in the agricultural sector, and the hunting ban, alongside the growing national and international emphasis on conservation and the preservation of endangered species, all coalesced to spark the beginning of a conservation boom. As some settlers embraced conservation and wildlife tourism out of economic necessity, others did so as a hobby and for lifestyle reasons. This was especially true of the fresh wave of white arrivants in the region who acquired citizenship and land on which they could live out their colonial imaginaries or position themselves as environmental missionaries and saviours of Africa (Nelson 2003).

Private Conservancies Go Global

As settler ranches changed hands, morphed into conservancies, expanded, and attracted international clients, the international conservation community took notice and looked upon large private landholdings in Laikipia as a laboratory for testing a range of innovative, value-added conservation activities. By the mid-1990s, converted ranches across Laikipia were being widely praised by conservation organizations around the world for their unique role in protecting Kenya's wildlife outside national parks – especially endangered and threatened mammals, including African wild dogs, black and white rhinos, elephants, Grévy's zebras, and reticulated giraffes. Novel sources of finance and experimentation with innovative, value-added activities soon began to take place in the region, ranging from dedicating properties to scientific conservation research, such as Mpala Research Centre, to arranging camel, horse, and bike safaris across the growing network of private conservancies in Laikipia.

Laikipia's privatized conservation laboratory was made possible by, and further attracted, renowned international conservation organizations, such as the AWF, Fauna and Flora International (FFI), and The Nature Conservancy (TNC), who entered into the new scramble for conservation land. In some cases, these organizations partnered with white landowners, providing finance and support for conservation activities and tourism ventures. In other cases, these organizations bought up former ranches, establishing new conservancies, research centres, and tourism facilities. As high-profile conservation organizations entered the scene, white landowners were no longer acting alone in transforming their ranches into wildernesses. They had the financial, moral, and political support of the international community.

Once again, Lewa led the way in establishing international partnerships with conservation organizations. From the start, Lewa was founded as a non-profit organization, which allowed the conservancy to qualify for funds on a charitable basis from international organizations ranging from the American Association of Zoo Keepers to the US Fish and Wildlife Service to the World Bank. In the early 2000s, when there were fears among settlers that private land rights could be curtailed in the lead up to Kenya's new Constitution and Land Act, Craig reached out to TNC. TNC specializes in purchasing land for conservation purposes and has a history of pursuing acquisitions that result in alienation, displacement, and highly exclusive enclosures (Laltaika and Askew 2021). TNC raised US$30 million to assist Lewa in restructuring itself in a way that strengthened its land tenure security – establishing a private company in Kenya to purchase Lewa before transferring use rights back to conservancy management (Van den Akker 2016). Along with a growing list of international donors, TNC continues to provide Lewa with significant financial support.

Since first working with Lewa, TNC has used similar tactics several other times in the region to help secure settler land for conservation. For example, when an old settler ranch in the north of Laikipia went up for sale in 2014, TNC outbid land developers who expressed interest in the property for subdivision and residential development. After purchasing the land, TNC transferred ownership to a Kenyan-based Trust – of which Ian Craig's son, Batian, is a trustee – and Loisaba Conservancy was established. In a press release, TNC affectionately referred to the new conservancy as "Lewa 2.0" (TNC, n.d.). This 56,000 acre conservancy provides secure habitat for elephants, lions, wild dogs, and an abundance of other species, and hosts several tourism ventures. Soon, however, the ranch will take its next step with rewilding by reintroducing rhinos.

FFI entered the private conservancy movement in Laikipia by adopting a slightly different approach. In 2003, FFI purchased one of the oldest settler ranches in Laikipia – previously owned by Lord Delamere. FFI made the acquisition with the financial backing of the Arcus Foundation, the private international philanthropic organization of Jon Stryker. Still a working ranch at the time of purchase with just a small, enclosed wildlife sanctuary called Sweetwaters, the property clearly had potential for rewilding, conservation, and wildlife tourism. The land had relatively high annual rainfall for the area, year-round rivers fed by streams on Mount Kenya, wetlands, and ecosystems far more diverse than those found in surrounding areas. FFI set about converting the 90,000 acre ranch into what is now Ol Pejeta Conservancy. The organization initially contracted Lewa to help set up the conservancy and involved both Ian and Batian Craig in Ol Pejeta's management, even though the two conservancies operate independently now. Within just a couple decades, Ol Pejeta became one of Kenya's most visited wildlife conservancies with 85,000 visitors a year – generating millions in income and winning numerous awards for its innovative approach to enterprise-driven conservation (OPC 2021a).

Rewilding Ranches and Conservancies

Rewilding settler ranches was not simply a matter of rebranding and redevelopment. To be recognized as a conservancy, the land needed to be seen as having the potential to support a wide range of species, especially endangered species and charismatic megafauna desired by conservationists and tourists. In short, entering the conservation business was not just an exercise in repurposing land, altering ranch management plans, and securing investments; fundamentally, it was about creating new ecologies. Settlers needed to create ecological conditions and relations that wildlife authorities, conservation organizations, and nature enthusiasts understood as natural and interpreted as wild. Although some large-scale farms and ranches had ideal conditions for rewilding – especially those with growing numbers of prey species, such as dik-diks, duikers, and other small *Bovidae*, as well as guinea fowls, hares, and warthogs – attracting and sustaining apex predators, such as leopards, lions, and spotted hyenas, and large herbivores, such as buffalos, elephants, and rhinos, was an arduous, costly, and time-consuming endeavour.

Before offering insights into the costs, expertise, infrastructure, people, and species involved in rewilding, a few more clarifications are needed about our use of the term. Quite a large body of natural and social science literature has emerged around the subject of rewilding over the years (Pettorelli, Durant, and Du Toit 2019), leading to a proliferation of meanings and applications of the term in different geographical contexts (Johns 2019). Some even understand the reintroduction of long-extinct species, such as woolly mammoths, within their historical ranges as an aspect of rewilding (Seddon et al. 2014). Although our understanding of rewilding is informed by scholarly insights from this literature, we avoid entering debates about the core meanings and applications, strategies and techniques, and temporalities that bound the concept (see Pettorelli, Durant, and Du Toit 2019). Instead, we adopt a simple definition of rewilding as any initiative that aims to restore a certain aspect of nature (Jørgensen 2015) – be it a landscape, ecosystem function, or species that is perceived by some humans to be missing yet needed.

Departing from uncritical definitions that suggest rewilding is about returning to a wilder or more natural state (for examples, see Jørgensen 2015), we maintain that rewilding is a hegemonic project informed by subjective judgments about which natures are worthy of restoration and which need to be eliminated or foreclosed to make this possible. There is an idea that rewilding initiatives operate with relatively minimal intervention: effectively, land is given back to nature. However, as we show, rewilding also has more active components, such as reintroduction, which refers to "the intentional movement and release of an organism inside its indigenous range from which it has disappeared"

(IUCN-SCC 2012, para. 2). Reintroduction has been gaining prominence as a rewilding approach in recent years, due in part to growing recognition that staving off biodiversity loss and species decline on a global scale requires pro-active, hands-on intervention.

In Laikipia, the rewilding that began to occur on settler land following the hunting ban emphasized the reintroduction of depleted, dispersed, and locally extinct species whose intrinsic value had come to be seen as desirable by settler ecologists for both economic and sentimental reasons. Transforming working ranches into habitats for such species required considerable re-engineering and constant intervention, which is another reason rewilding should not be mistaken as a return to a prior era when it is often alleged that nature and humanity existed independently from each other. Due to the capital, expertise, and other resources needed to re-engineer settler properties and constantly intervene in restoring and maintaining conservation landscapes, it was during the rewilding era that the social category of settler ecologist, as we have described it, began to diversify and expand.

From the late 1970s onwards, white landowners undertook both large-scale rewilding initiatives and focused reintroduction projects. The first step in rewilding settler ranches generally involved reducing livestock herds or removing livestock from the landscape altogether. Removing large herds of livestock quickly was an efficient and effective way to reduce grazing pressure and allow vegetation, including grasses, shrubs, and trees, to regenerate. As ranches were partly destocked, internal livestock fences were also removed from properties to create open, uninterrupted landscapes of savannah and shrubland. During this same time, some ranches also reduced or gave up other activities in their livelihood portfolios, such as gardening and fish farming, which increased water volumes on ranches and transformed habitats, such as wetlands which had been drained or reduced in previous years. Livestock dams also became important sources of water for wildlife in arid and semi-arid areas, as, according to some, wild animals felt safer drinking from dams less frequented by livestock and herders.

Some private properties also began to construct new water infrastructure on their land to attract wildlife to desirable habitats and other strategic locations. This remains a common activity today. Many ranches are constantly upgrading water pumps, constructing new dams, and running new pipes that draw from the Ewaso Ng'iro or other water sources fed by the glaciers and streams on Mount Kenya. For example, Ol Pejeta built pipelines across the conservancy,

Photo 2.1. Elephants forage around a permanent water source in front of tents at Sweetwaters Serena Camp

along with multiple artificial water holes, to keep key species on the property even during dry season. At Sweetwaters Serena Camp, luxury tents have been constructed in a semi-circle around a permanently supplied water hole set directly in front of the main dining area so that wildlife is attracted to and spends significant time at the camp – particularly during periods of drought (see photo 2.1). Recent studies state that the abundance of artificial water holes on Ol Pejeta, along with undesired encounters with humans outside the conservancy, lead species such as elephants to limit the distance they travel and time they spend away from the conservancy (Sernert 2017). In addition to attracting and retaining wildlife inside conservancies, artificial water provisioning is often used to concentrate wildlife in prime viewing areas (Sutherland, Ndlovu, and Pérez-Rodríguez 2018).

As ecological conditions on settler ranches and conservancies have been transformed, many previously absent species began to flourish in the landscape. In fact, during the 1980s and 1990s, wildlife abundance decreased almost everywhere in Kenya aside from Laikipia, where most populations were stable or growing (Kinnaird and O'Brien 2012). The elephant population increased from 2,969 to 6,365 between 1992 and 2012 – a 114 per cent increase in just twenty years (Litoroh et al. 2012). As part of the greater Samburu-Laikipia Ecosystem, Laikipia came to boast Kenya's largest elephant population outside national parks. Growing wildlife populations in the area were mainly attributed to what we describe as the rewilding of settler landholdings (Thouless and Sakwa 1995; Georgiadis 1997; Mizutani 1999).

During the early years of rewilding on Lewa Wildlife Conservancy alone, zebras increased from 374 to 1,529 individuals; giraffes from 181 to 534; and impalas from 231 to 560 (Kock 1995). Even some species that were highly endangered, if not locally extinct, were attracted back to the region by conservancies. For example, for years, wild dog numbers have been steadily declining across most of their historic range on the continent. Yet, with a consistent supply of prey readily available across a growing network of wildlife-friendly properties, they were attracted to conservancies such as Lewa where their numbers are growing to this day. Over the past two decades, the resident wild dog population on Ol Pejeta has also grown from zero to thirty-two individuals spread across two packs known as Bahati and Ol Pejeta. In 2021, the two units produced eighteen pups (OPC, n.d.-c). Ol Pejeta attributes this population growth to suitable habitat and prey in the conservancy and strategic fences that enable the dogs to roam across the wider landscape using secure corridors. As conservancy management explains, "the majority of our wildlife is free to migrate in and out … as they wish [but] most of them decide to stay inside our fences because we have created a safe haven for them" (OPC 2019).

Although species such as wild dogs were naturally attracted to conservancies where prey was abundant and readily available, reintroducing others that were less inclined to spend time in former ranches was far more difficult. Initially, these efforts were also quite haphazard and at times involved questionable methods. For example, it has been reported that some ranches built up their wildlife numbers in the 1970s by capturing "animals under the false pretence that they were ill and needed treatment," before relocating them to their own properties (Van den Akker 2016, 127). In 1978, wildlife authorities became involved in a situation where elephants had destroyed numerous crops belonging to smallholders in central Laikipia. In response, aircraft, vehicles, and a line of four hundred men attempted to drive the large herd into nearby ranches (Van den Akker 2016). Over time, these rudimentary approaches evolved into more coordinated measures with wildlife authorities translocating animals to settler ranches and conservancies in a more systematic way. However, some of the same tactics, such as herding large animals away from communities and into private conservancies with light aircraft, are still used to this day.

As discussed earlier, rhinos became the focus of early reintroduction efforts, with the first rhinos moved to Solio in the 1970s. Before long, starter black rhino populations were also translocated to Lewa and Ol Pejeta. By most accounts, rhinos were locally extinct in Laikipia by the 1970s. However, by the late 1980s, numbers were increasing by 10 per cent a year as a result of successful breeding (Cohn 1988). Today, Ol Pejeta claims to host the largest black rhino population in East Africa, as the result of a series of translocations and post-release breeding, which expanded from 46 individuals in 2004 to 140 individuals in 2021.

Beyond rhinos, other less-known species were also reintroduced to settler ranches and conservancies. In the 1990s, Lewa Wildlife Conservancy gained a starter population of Grévy's zebras, which are endemic to northern Kenya and some areas in Ethiopia. The species had been heavily hunted for the unique and striking patterns on its skin, which was in high international demand. The idea was that Lewa represented an ideal, secure environment in which this once-coveted trophy species could recover. As the population grew, eight of the individuals were later moved to Ol Pejeta to live in a predator-proof enclosure.

Securing Rewilded Ranches and Conservancies

As more species of wildlife began to repopulate more ranches and conservancies through rewilding, settler ecologists also felt a need for better security and stricter surveillance to protect endangered species and wild assets. Thus, demand for electrified perimeter fencing, sometimes accompanied by special cattle ditches, skyrocketed. The fences had to be specially designed to allow some wild species, such as wild dogs, to come and go from conservancies while constraining the movements of others, such as rhinos. Fences also needed to stop mobile herders and cattle from accessing land areas they or their ancestors might have accessed in the past. By the late 1990s, dozens of private ranches and conservancies had built expensive, state-of-the-art fences to protect resident wildlife populations (Thouless and Sakwa 1995). In many cases, these fences were not simply electrified, but also connected to alarms that notified and activated an armed response team when the current dropped (Thouless and Sakwa 1995). In this way, the rewilding of settler ranches can also be understood as the privatization of fortress conservation, which, as discussed above, had also become a defining feature of the national park solution. As described by Ian Craig, Lewa developed "extensive security in the form of fences with a heavily armed security force to protect our wildlife. This effectively isolated us from everything outside the fence" (Craig as quoted by Thomas 2000, 48).

Importantly, it was not just the movements of vehicles, people, and livestock that were increasingly constrained by vast networks of electrified fences springing up across the plateau. Wild species started to be heavily monitored and afforded fewer options when it came to navigating the landscape outside conservancies. Wildlife passages – or corridors – in and out of conservancies tended to be placed in areas deemed appropriate and safe for migratory species to pass through. As mentioned, these fence gaps were also designed to block endangered species, such as black rhinos, from roaming beyond private fortress conservation areas. Throughout the 1990s and 2000s, the design of conservancy fences and fence gaps was adapted and refined until wildlife were left with few options in terms of where they could move and how they could get there. More recently, the positioning of fence gaps and securing of wildlife corridors has become a coordinated nationwide effort now targeting the entire Samburu-Laikipia Ecosystem.

Similarly, predator-proof fences have been used within conservancies to protect rare and endangered herbivores from over-predation. As populations of predatory species in conservancies steadily increased with rewilding, they became yet another threat to rare and endangered herbivores that had been reintroduced in conservancies. This became a serious issue for Grévy's zebras and Laikipia hartebeests at Ol Pejeta. In response, the conservancy designed special predator-proof enclosures to help protect key breeding populations. Here, it is also worth noting that security teams and wildlife rangers are required to report on any ill, injured, abandoned, or orphaned animals they encounter during patrols. Depending on the situation and the species, mobile veterinary teams may be called in to treat the animal on site or transfer it to a sanctuary for further care.

In addition to the proliferation of electric fences, rewilding led to a surge in demand from private ranches and conservancies for wildlife rangers, security personnel, military grade weapons, and surveillance technologies. Security teams on private ranches and conservancies operate 24/7 and, depending on the property, are responsible for aircraft surveillance, guarding entrances and patrolling perimeters, handling sniffer dogs, and offering hands-on protection to rhinos or other critically endangered species (see photo 2.2). With each passing year, surveillance systems and specialized paramilitary teams have become better equipped to respond to perceived threats to rewilding

Photo 2.2. Security fence around endangered species enclosure at Ol Pejeta Conservancy

efforts. For example, one writer who was granted access to Lewa's monitoring operations room likened the facility to NASA's mission control room:

> I'm sat in a space resembling mission control at NASA. It's Lewa's 24/7 monitoring operations room, with blinking lights, satellite imagery and short radio messages beeping away, plus a phone ringing in the background. The walls are covered by giant screens depicting various different maps of the conservancy, each of which is covered in dots and lines indicating the location of different collared animals. (Fitch 2016, para. 13)

Investment in the latest, cutting-edge security tactics and technologies has become a marker of a strong performing conservancy. Many of the conservancies backed by international conservation finance devote a good deal of their annual reporting to information about upgrades to security systems and teams, as well as statistics demonstrating the success of their interventions. They also regularly profile security teams and wildlife rangers on social media platforms.

In stories about the transformation of ranches into habitats for some of the world's rarest and most endangered species, white landowners are often described as "catalysts," "pioneers," and "role models" of innovative approaches to restoring and saving African nature. Yet, as this section demonstrates, the rewilding of settler landholdings should not be mistaken as a return to a bygone era when Laikipia was teeming with wildlife, devoid of people and in a pristine state of nature. In practice, rewilding has been shaped significantly by ecological changes settlers had already wrought upon the landscape during earlier periods of colonial settlement, as well as a series of crises that caused them to seek out new ways of maintaining their foothold in the landscape through wildlife. The approaches they took to rewilding manifested in engineered landscapes requiring constant intervention to attract and retain certain desired species and maintain ecological aesthetics that aligned with colonial imaginaries of Africa. As Van den Akker writes, the private conservancy was "a 'new' nature that did not derive its credibility from the historical absence of human interference, but from deliberate landscape engineering" (2016, 127). Although the landscape of Laikipia may now be presented as primordial, wild, and natural, settler ecologists have been instrumental in its making.

Making Settler Ecologies: Rewilding

In April 2015, we are sat in the back seat of a four-wheel-drive Toyota Hilux that is powering its way through wet black cotton soil as it proceeds out of a settler ranch located at the heart of the Laikipia Plateau. The soil is loose and slippery and will soon become all but impassable if the rains continue. The vehicle is being driven by Miles, a white Kenyan, who is accompanied by his Meru wife, Eleanor, in the front passenger seat.

We slept over at the ranch after spending the day in conversation with Miles, flipping through his photo albums and listening to countless tales accumulated from a life defined by wildlife. Miles was born and raised in Kenya. His father was a white hunter who operated safaris for high-end clients, including members of the Las Vegas Rat Pack. It was perhaps inevitable that Miles would shoulder a gun himself. Miles led a special anti-poaching unit when KWS was under the leadership of Richard Leakey, who was notorious for his aggressive stance against suspected poachers. Eventually, Miles abandoned the gun and followed in his father's footsteps by setting up his own safari ventures and ecolodges, some of which "went bust," he admits. Miles also became a selfproclaimed "animal whisperer," rescuing and rehabilitating animals such as mongooses, genets, and cheetahs. He claims to have a special bond with wildlife, including dangerous animals like elephants, and has a stockpile of photos that serve as evidence of this.

It had been a long twenty-four hours, and we are silent as the main road outside the ranch comes into view. However, the tail end of an observation being shared by Miles grabs our attention. Wagging his thumb at the back seat, Miles is explaining, "Us white people, we love wildlife, can't get enough of 'em." He then points his thumb in Eleanor's direction and states matter-of-factly, "Africans don't give two cents about wildlife. I don't mean that in a bad way, I'm married to one. It's just a fact." We look at Eleanor, who is smiling, dressed in a khaki shirt and camouflage trousers. "It's true," she says with a single affirming nod. Miles is not the first white Kenyan to share this sentiment with us. The stereotype is fairly common among this segment of Kenya's white community, who sometimes seem to strategically forget their own tenuous history with wildlife in this part of Kenya.

The notion that white Kenyans have a unique love of wildlife – and that rewilding was primarily motivated by this love – circulates well beyond the white community in Laikipia. About one month after Miles shared the sentiment with us, we found ourselves sitting in Marina Cyber Café in Nanyuki, about 40 km southeast of where we visited Miles and Eleanor at the ranch. Presently, we are in the middle of an interview with one man who has worked for Il Ng'wesi, Northern Rangelands Trust, and, most recently, Laikipia Forum. The rainy season is still going strong and the dull, roaring sound of a deluge impedes our conversation. One of us inches our recording device down the table nearer the speaker, placing it between an empty Coca-Cola bottle and a small dish containing a now-cold samosa. The speaker clears his throat and leans forward, repeating himself.

"There is an expression, 'You don't know what you have until it is gone,'" he enunciates loudly towards the device. "I heard that, in Europe, US, and such places, they killed all wildlife. Now, those people come here to see our wildlife because they have none." The man pauses, unfolds, and refolds his hands, as though stating the obvious, and continues: "That is why white people love our wildlife so much, because now they have none." Since listening to Miles discuss the love white people have for wildlife, we have been relaying the sentiment to Maasai acquaintances. While contesting Miles's claim that "Africans" have no love for wildlife, they often explain settlers' love of wildlife by reciting a version of the famous verse from Joni Mitchell's "Big Yellow Taxi," in which paradise is paved to make way for a parking lot.

The "Big Yellow Taxi" theory resonates with us. As we show in this chapter, settler ecologists' rewilding practices and love for wildlife were born out of a fear of losing paradise. Events in the twenty-first century created multiple material and metaphysical crises for settlers. These crises were linked to demands for land reform in Laikipia and the anti-colonial struggle of the Land and Freedom Army in the build up to independence, as well as changes to agricultural and economic policies and the hunting ban that followed formal independence from Britain. All of these events had destabilizing effects on settler-dominated industries and socialities. Rewilding was born out of the settler imperative to claim back paradise: to secure their land, diversify their livelihoods, and improve their chances of economic survival in post-colonial Kenya. Through rewilding, settlers identified and seized hold of a new "political economy of belonging" (Pailey 2021) that gave them staying power.

Given the motivations behind this phase of ecological transformation, rewilding had little to do with reintroducing and revitalizing ecologies that *actually* existed prior to colonial settlement. At this point in time, settlers had no utility or affinity for the indigenous ecologies that dominated Laikipia prior to colonial settlement. The densities, distributions, and diversities of animals and plants that settlers placed on their properties during this period aligned with colonial imaginaries of African wilderness as unspoilt Gardens of Eden teeming with charismatic species but devoid of humans and their activities (Neumann 1998; Singh and van Houtum 2002; Ramutsindela 2004; Garland 2008). These rewilded landscapes provided settlers with access to capital and permission to secure their property in the name of protecting "African wilderness."

Importantly, international conservation organizations have been among the biggest allies and supporters of white Kenyans involved in rewildling in Laikipia and have played a crucial role in helping settlers resolve many of the crises they faced in the post-independence era. Although many white Kenyans in Laikipia had experience managing ranches, few had the capacity, resources, or knowledge needed to transform their ranches into flourishing private protected areas. International conservation organizations, on the other hand, had ready-made access to the financial and human capital and technical skill required for rewilding, including financing arrangements, funding networks, legal teams, research connections, and partnerships with state wildlife authorities. They have worked closely with white settlers to replicate and recreate national park operations in former ranches by copying and pasting land use planning, resource management, veterinary services, and anti-poaching and security operations typical of national parks onto private land. This era of rewilding marked the beginning of the international conservation sector's role in producing settler ecologies in Laikipia and preserving private paradises for settlers and wildlife alike.

3 Repeopling

In early 2015, northern Kenya remains in the grip of prolonged drought. The air is dry, hot to breathe, and filled with dust. Sitting in a hired four-wheel-drive Suzuki, we turn off Nanyuki-Rumuruti Road onto grated murram, bouncing and meandering our way northward into the ranching heartland of Laikipia. We will soon pass settler ranches and conservancies such as Lolldaiga, Ol Jogi, Mogwooni, and El Karama. Further afield are Segera, Sosian, Loisaba, Mugie, and others. The windows of the small Suzuki are open just a crack as the heat of the car is preferable to the dust kicked up by lorries trundling along the road, which carry sand mined from riverbeds to Nanyuki and other distant urban centres.

As we drive, a makeshift shelter on the roadside becomes visible. The shelter is made of carrier bags tied to a fence wire in the fashion of a lean-to. We then pass another makeshift lean-to and later another, all stained the colour of dust and tied like clothes on a line to fences so long they stretch out of sight. These shelters belong to herders migrating their famished cattle along the roadway, where sparse clumps of brown grass between the road and the fences are free for the taking. On the other side of the fences, the vast rangelands of some of Kenya's most elite conservancies appear empty and idle, except for a few giraffes and zebras leisurely grazing in the distance.

The deeper into the plateau we drive, the more cattle begin to appear around the makeshift shelters. The cattle are in bad shape: their hides look like bed sheets draped over skeletons that were possibly once cows. The animals huddle together along the fencing or in small *boma* consisting of a few branches of whistling thorn acacia, as if competing for shade invisible to the human eye. The nearby herders also look weary, propped up on one arm under their shelters and squatted under thorny bushes along the road.

Sitting in the Suzuki with us is a recent acquaintance, Daniel, a member of the Il Ng'wesi community who works as a guide at the eco-lodge where we first met him. The man, in his late thirties, is also a pastoralist and keeps his livestock at Chumvi, near Borana and Lewa. The more shelters, herders, and cattle we pass, the more we all discuss their situation. These herders are from arid regions further north in Samburu County. For generations, they and their ancestors have come to the plateau to access areas of pasture specially reserved for hard times like this. However, in the past, herders would have been able to move about Laikipia with greater freedom without wildlife conservancies limiting their movement. Our acquaintance looks back at one herder as we continue our drive and says, "If that is what it means to be a pastoralist, I don't want to be a pastoralist."

Fast forward to April 2017. Heavy rains have just arrived this week following another long, punishing drought. We are driving through the same area, only this time we are travelling at night. The windows are down to allow the air to flow in and cool the vehicle. There is tension in the air. Some nearby conservancies were recently invaded[1] by migrating herders and their cattle. Most herders were simply interested in leading their cattle to pasture and water in conservancies. However, conflicts with conservancy personnel and security teams also occurred. Reports from conservancies claimed that herders harassed tourists, burned tourism facilities, and killed wildlife. White settlers also became involved in the conflict. On 6 March, a white lodge owner and manager, Tristan Voorspuy, was found dead on Sosian Game Ranch – allegedly shot and killed, along with his horse, by herders when he approached them with a gun on horseback. On 22 April, another settler, Kuki Gallman, was shot and injured while patrolling her ranch, Ol Ari Nyiro, with wildlife rangers.

We fail to see a patch of black cotton soil in the darkness of the night and the vehicle comes to a lurching stop. In an attempt to escape the inevitable, Daniel shifts the car into reverse. The wheels spin and the

---

1 Like "poacher," the terms "invasion" and "invader" are politically loaded and often used to demonize entire groups of people (i.e., pastoralists) who have temporarily or indefinitely opted to disregard the boundaries of conservation areas. For this reason, some, such as Bilal Butt (2014), refer to "incursions," recognizing that the movement of people, livestock, and even wildlife in and out of conservation areas can be understood as a form of political agency. Nevertheless, we use the term "invasion" in this chapter to give readers a sense of the violent discourse that surrounded pastoralist incursions in Laikipia during 2015 and 2017.

engine revs, but the Suzuki does not move. We are stuck. We get out of the vehicle to survey the situation and quickly come to the conclusion that we are going to have to call for help. After heavy rains like this, black cotton soil is slick when a vehicle is moving, only to transform into a deep, heavy mud as soon as the movement stops. We make a call and begin a long wait, stretching our legs by walking circles around the car while gazing at the stars suspended over the plateau.

In the distance, we notice a short line of waist-high lights bobbing up and down across the darkened plains. We discuss what they are – at first mistaking them for glow-worms. Then, we hear several rapid warning shots fired in the distance and realize our mistake. The lights are torches held by herders making their way onto private land to graze their cattle under cover of darkness. The torch lights switch off. The shooting stops. We get back in the car to continue our wait. We look out across the landscape for signs of an injured party or further conflict, but it is eerily silent aside from the sounds of crickets, frogs, and toads thankful for the recent rains. Eventually, our help arrives and the car is freed, covering us all in thick, black mud upon its release. We drive away, hopeful that the herders somewhere nearby are safe as they attempt to graze their cattle on the only land in the area with grass left under the cloak of darkness.

The invasion of ranches and conservancies in Laikipia during 2017 marks an important moment in time. These invasions occurred in the aftermath of recurring droughts that took a serious toll on pastoralists and their livestock across the region. However, they also escalated in the build up to a national election in August 2017. Some members of the opposition party seized the drought as an opportunity to secure votes from pastoralist constituencies, where voter turnout is historically low, by declaring war on settlers and vowing to dismantle white-owned properties if elected (Dahir 2017; Haggard 2017). A member of parliament from Laikipia North was eventually implicated in the murder of Tristan Voorspuy – although charges were later dropped – and a bullet used by herders to kill one ranch worker was traced to a government ordinance factory (Ndirangu 2017; Ebru News 2017). Political encouragement and support for the invasions led some settlers to question their future in the country. As one conservancy owner said to us in April 2017, "if we can just get through the election, hopefully things will calm down."

The concerns of white Kenyans began to dissipate as the election neared and government bodies rallied in support of settlers and conservancies. In a remarkable display of loyalty to the wildlife sector, seen as a major source of employment and revenue in Laikipia, the government initiated military operations mandated to root out and disarm herders trespassing on white-owned land (Agutu 2017). These operations were often heavy handed and punitive. For example, in Ol Moran Ward, police and military personnel combined forces to massacre over 350 cattle belonging to herders that had migrated to Laikipia from Samburu (NTV Kenya 2017). The message to herders and indeed the global conservation community was clear: the government was on the side of wildlife, meaning it was also on the side of settler ecologists.

In this chapter, we reflect on how these types of events forced settler ecologists to reorganize ecological relations in Laikipia to resolve and make productive long-standing tensions between pastoralism and colonial settlement. We argue that to deal with deep-seated and recurrent tensions with pastoralists that were heightened through rewilding, settler ecologists have begun to incorporate pastoralists and aspects of pastoralism into settler ecologies. As a result of these efforts, it has become increasingly common to see pastoralists, livestock, and wildlife sharing space in Laikipia. Yet, rather than fully opening up settler ecologies to pastoralists and their ecological knowledge, "repeopleing" efforts tend to fix pastoralists in instrumental and inequitable positions that both maintain and advance settler ecologies. Ultimately, this chapter illustrates how inclusion in settler ecologies – much like exclusion and dispossession – can also be violent, as has been argued in relation to Indigenous inclusion in other contexts by scholars such as Tania Li (2014) and Kēhaulani Kauanui (2017).

The chapter is organized around two different approaches that have been designed and used to incorporate pastoralists into settler ecologies since the 1990s. In the first half of the chapter, we look at how some private conservancies have incorporated pastoralists and pastoralism into their properties with certain conditions. We show that in addition to helping mitigate conflict, this move also has wider ecological, social, and economic benefits for private conservancies. In the second half of the chapter, we consider efforts led by settler ecologists to transform community land into conservancies, where space is allocated for pastoralism, wildlife conservation, and eco-tourism. This move benefits wildlife by securing pastoralists' rangelands for biodiversity conservation while at the same time prescribing pastoralists with new roles and rules on their own land. Ultimately, in both scenarios, the inclusion – rather than exclusion – of pastoralists and pastoralism is used to transform the ecological aesthetic of conservation landscapes in ways that naturalize ecological and socio-economic relations of value to settler ecologists.

Incorporating Pastoralists into Settler Ecologies

For decades, efforts to create and maintain an ecological aesthetic that appeals to elite tourists have involved the pursuit of guaranteed wilderness encounters and experiences. In Laikipia, these guarantees usually include an exclusive safari experience that provides unencumbered views of fauna and flora and unfettered access to pristine wilderness without sacrificing luxurious amenities and service (Ávila-García and Sánchez 2012). The promise of exclusivity is especially important to elite clients, who wish to avoid waiting in a buffet queue or for a view of a leopard kill. High-end tourists demand to feel special and expect nothing less than the spectacular. They wish to be the only sign of human life for miles; to be offered gin and tonics on the bonnets of antique Land Rovers at sundown; to be served pâté in front of an open fire after an evening drive, wrapped in a Maasai *shuka* for added warmth; to lounge in pools overlooking watering holes frequented by elephants, giraffes, and impala; to track lions and wild dogs on foot or by horseback; and to take selfies with the most endangered mammals on Earth. These types of tourists expect to touch down on private airstrips and helipads, climb into shaded safari vehicles, and be chauffeured through landscapes teeming with all forms of life except humans.

The Lewa-Borana Landscape encompasses all these ideals. The ten lodges nestled across the landscape offer luxury, tailor-made safari packages for elite guests with costs ranging from US$750 to over US$1,850 per night per person. As UK-based travel agent Expert Africa explains, "With a vast swathe of wilderness and few of the usual park rules," lodges in Borana and Lewa Wildlife Conservancies treat guests to "exclusive safari experiences" and "understated luxury" (Expert Africa, n.d., para. 1). Alongside traditional game drives in open-top four-wheel-drive vehicles, there are numerous other ways for visitors to encounter Kenya's wildlife and wilderness in the Lewa-Borana Landscape, be it by horseback, mountain bike, guided trail runs, treetop canopy tours, fly fishing, wildlife tracking, paragliding, or lounging in infinity pools overlooking strategically placed water holes. With exclusive access to 91,000 acres, discerning travellers visit this landscape to view megafauna, including elephants, rhinos, big cats, buffalos, giraffes, and abundant birdlife, in comfort. Lodges in the Lewa-Borana Landscape are also widely promoted as an ideal option for those searching for luxury family-friendly safaris – partly due to the range of activities on offer, as well as a unique rock formation fabled to have inspired Pride Rock in Disney's animated film *The Lion King*.

In more recent years, patrons of exclusive upmarket wildlife areas – even those visiting the likes of Lewa and Borana – have adopted new expectations about the role of wildlife conservancies in wider society. Many tourists now hold strong beliefs around the responsibility of conservancies to invest in nearby communities and ensure local populations benefit from conservation and tourism. This may be because clients want to avoid wrestling with any residual feelings of guilt while trying to enjoy their holidays, or because they see their patronage as an act of conservation if it enhances local buy-in and support for conservation. In Laikipia, these expectations became particularly pronounced following national and global media attention on the conflicts of 2017.

Accordingly, conservancies are going to greater pains to explain and publicize how they contribute to local livelihoods and benefit the lives of individuals in nearby communities on their social media accounts, websites, and other promotional materials.

These shifting expectations have made certain highly controlled and choreographed encounters with humans a more anticipated and normal part of the safari experience in Laikipia in recent years. For example, visitors to Lewa, Borana, and other high-end conservancies are usually offered the opportunity to interact with pastoralists outside conservancies as part of their safari experience. As Borana Conservancy explains to potential guests, "We encourage all the guests to Borana Conservancy to visit our neighbouring Maasai community for a cross-cultural experience – dancing, shooting bows and arrows, drinking tea and trying your hand at beading" (BC, n.d.-b, 16). The conservancy's website also explains that Borana has a close relationship with one local community north of the conservancy (Il Ng'wesi) that is keen to welcome visitors. Tours can be organized where a "Maasai elder will give you a tour of the village and describe the history of the Maasai tribe. Various members of the community will show you their customary ways of life and traditions. The tour usually ends with a traditional tribal dance which you are encouraged to take part in" (BC, n.d.-b, para. 28–9). These cultural activities are also offered by other conservancies, such as Lewa, Loisaba, and Ol Jogi, and represent one way that pastoralists are incorporated into commodified tourism encounters afforded by settler ecologies.

Additionally, conservancies and lodges are rarely shy about increased efforts to incorporate individuals from pastoral communities into their operations as guides, lodge staff – including cleaners, cooks, and servers – and rangers. Most of the safari guides we have interacted with over the years who work in the Lewa-Borana Landscape grew up in and around Il Ng'wesi. Some of them attended schools supported by one of the conservancies and studied at Utalii College in Nairobi with scholarships funded by conservancy donors and guests. This is true of many who occupy other skilled positions in conservancies and lodges in accounting, human resources, and public relations – although most lodge managers in private conservancies still tend to be white Kenyans or foreign nationals. Many of the security guards and rangers hired to protect tourism facilities and wildlife in conservancies are also from nearby communities. Lewa describes its anti-poaching units and wildlife rangers as "hand-picked local militia" that include ex-poachers looking to "atone for [their] actions" (LWC 2017).

These forms of inclusion satisfy expectations among conservancy guests, donors, and investors alike, allowing elite and wealthy clients to rest assured knowing their exorbitant fees are being used to do good. Yet including pastoral communities in conservancy life also serves other interests of settler ecologists. For instance, creating employment opportunities and supporting community initiatives shores up support for settler ecologies and hedges against future hostility towards an exclusive conservation sector with an insatiable appetite for land. In this regard, conservancy investments in education are especially important. By building schools, developing conservation-related curriculum – including free field trips into conservancies – and funding bursaries and scholarships for all levels of study, conservancies are investing in future settler ecologists. This new generation of settler ecologists may not resemble white Kenyans or foreign nationals, but they are being equipped with a shared knowledge, language, and skillset about what ecological relations have value and nurtured to maintain settler ecologies in various ways.

Incorporating Pastoralism into Settler Ecologies

Efforts to make conservancies more inclusive of nearby communities have also extended to the more-than-human realm, giving indigenous cattle a foothold in some of Laikipia's most elite tourism spaces. In recent years, droughts, invasions, and conflicts have sparked coordinated initiatives among conservancies to grant a select number of cattle from some pastoralist communities with temporary access to private conservancy land. The specifics of these arrangements differ from one conservancy to the next, but, across the board, permitting access to conservancies for livestock aims to alleviate longstanding tensions between settlers and pastoralists. After all, many elders in places such as Il Ng'wesi still recall a time when no fences existed between their community and the settler ranches they have watched grow into world-renowned conservancies. The incorporation of pastoralism into settler ecologies represents a kind of lifeline settlers throw to their neighbours.

Photo 3.1. Livestock grazing on a private conservancy as part of an LTM initiative

The primary means through which pastoralists' cattle have been included in settler ecologies is through livestock-to-market (LTM) initiatives. The first formalized LTM programs in the region, called Northern Rangelands Trust-Trading (NRT-T), began in 2006 as a partnership between two private conservancies, Lewa Wildlife Conservancy and Ol Pejeta Conservancy, and a community conservancy organization called Northern Rangelands Trust (NRT). By 2020, several other private ranches and conservancies were participating in this initiative, including El Karama, Loisaba, and Mugie. The stated ambition of NRT-T is to provide pastoralists in NRT conservancies with a consistent, reliable, and fair market for their cattle. To achieve this, NRT-T purchases cattle from people in community conservancies and transfers them to settler properties where they are inoculated, quarantined, and fattened (see photo 3.1). Somewhere between four and eighteen months after purchase, all NRT-T cattle are taken to Ol Pejeta where they are slaughtered at the ranch's industrial abattoir before being sold at markets in Nairobi and abroad.

Other settler ranches and conservancies in the region have developed similar programs. Borana Conservancy, for example, launched its own

LTM initiative in 2016. Unlike NRT-T, Borana does not buy the livestock. Instead, it provides pasture and day-to-day grazing coordination, along with marketing support, for pastoralists from nearby community conservancies who elect to graze their livestock on Borana's land – usually during times of drought or other hardship. On an annual basis, dates are announced for entry into the program. On these dates, pastoralists bring their cattle to Borana's gates and hand them over to Borana ranch hands for care. While at Borana, these livestock are fattened and treated for ticks, infections, and disease. Once the cattle reach a weight and level of health appropriate for sale, Borana helps identify a buyer for the cattle and negotiates a fair price. Borana receives a 20 per cent cut on the sale as a grazing management fee while pastoralists reportedly receive higher prices for healthier cattle.

LTM initiatives are promised to benefit pastoralists in several ways. The main selling feature is improved market access, partly under the assumption that pastoralists overstock their herds because they lack access to good, reliable markets that would allow them to sell. Initiatives like NRT-T also claim to help pastoralists earn higher prices for livestock than traditional markets because they produce healthier livestock with higher quality meat. By helping pastoralists increase their income through livestock sales, LTM initiatives aim to reduce the number of livestock kept by pastoralists. The idea is that fewer livestock and smaller herds will improve rangeland conditions in community conservancies and leave more grass for wildlife. In turn, improved rangelands are meant to create more suitable habitats, increase biodiversity and wildlife, and generate more tourism revenues in community conservancies, further reducing the need for large herds of livestock to be kept in the first place.

Despite these ambitions and promises, the benefits of LTM initiatives are heavily debated by pastoralists and experts on pastoralism alike. To start, once a pastoralist surrenders their livestock to the LTM scheme, they no longer have a say over the price the livestock will eventually sell for. In some cases, participants in these schemes have reported that their livestock were unfairly priced. Furthermore, many communities fear that LTM schemes threaten their livelihoods and cultural heritage as they push for destocking on community lands to pave the way for more wildlife-centric landscapes. This fear is somewhat justified given the discourse surrounding NRT-T. For example, as a trustee of TNC, one of the largest donors of NRT-T, explains, "The ultimate goal is a shift to fewer, healthier cows" (TNC, n.d.). Finally, only pastoralists that adhere to NRT-imposed grazing, security, and management rules on conservancy land are deemed eligible to participate.

Although the extent to which LTM initiatives benefit pastoralists and pastoral ecologies is debatable, there is no debate as to whether these initiatives benefit settler ecologies. LTM initiatives are strategically used by settler ecologists to improve grazing quality for wildlife in private conservancies and ranches. This is a particularly important strategy among conservancies that have recently destocked their own livestock herds for rewilding purposes. LTM schemes provide conservancies with access to the nutrient-rich dung of small herds of cattle, which is essential to facilitating and maintaining healthy grassland ecosystems and ensuring a good mix of ground cover and palatable grasses for wildlife. By tightly bunching cattle together in pre-demarcated grazing areas and shifting livestock from one location to the next across conservancies, cattle belonging to pastoralists help improve the palatable forage for animals and attract wildlife conservancies without the need for conservancies to maintain their own larger herds (NRT 2019).

LTM initiatives also help to endear the wildlife sector to pastoralists. As discussed, historically, settlers have focused on securing the perimeters of their conservancies and ranches. However, the implementation of LTM schemes has opened up these same conservancies and ranches as potential sources of pasture for some pastoralists. The option to participate in an LTM scheme improves the overall perception of conservancies in the eyes of some pastoralists. LTM grazing arrangements have been especially impactful in winning pastoralists over during recurring drought. As it is common for pastoralists to receive little support from the government during crises, the ability to access pasture in conservancies without threat of punishment or death is well received. Some proponents of LTM initiatives have suggested that the grazing arrangements are so well received that pastoralists in neighbouring communities have begun to patrol conservancy borders themselves to ensure pasture and water reserves inside are secure (Fox 2018a). If this is true, LTM initiatives appear successful at enrolling some pastoralists in the work of protecting settler ecologies as well.

In a recent special feature in *Condé Nast Traveller* about Borana Conservancy, the writer describes the inclusion of pastoralists and pastoralism as part of the success story of elite conservancies:

> Originally a cattle ranch, like much of the land around here, it shares a boundary with Lewa Wildlife Conservancy, which has been at the forefront of rhino conservation in Kenya for more than two decades. Borana operates both as a working cattle ranch, traversed by the nomadic Maasai with their cows and goats, and as a wildlife sanctuary. It's a balancing act

that's anathema to safari purists who prefer even the illusion of pristine wilderness, devoid of any sign of human habitation but with the commitment and involvement of local communities it has proved a successful conservation model. (Browne 2017, para. 5)

Rather than denying or working to erase Indigenous presence in the landscape altogether as might have been done in the past, settler ecologists are figuring out how to strategically use certain aspects of this presence for their own purposes – using this presence to signify a progressive conservation model and gain local support.

Of course, great care is still taken to ensure that elite tourists encounter livestock only in certain contexts and with adequate warning and explanation. Even conservancies that double as working ranches need to maintain a reputation as wild spaces worthy of tourism fees and conservation finance. One morning, while we were having tea at a private lodge along the banks of the Ewaso Ng'iro, a miscommunication upset the carefully choreographed movements of livestock and people in this particular landscape. As we sat, we heard cow bells growing louder over the sound of the bubbling river and cooing doves, accompanied by the snorting and grunting of cattle. On the opposite banks of the river, cows slowly appeared one by one until the entire hillside was dotted with large black, brown, and white animals. The cows were ushered past at no great speed by a young herder wearing a red *shuka* and khaki shirt, chewing on the branch of a toothbrush tree (*Salvadora persica*) and checking the signal on his small Nokia as he followed the cows through the tall grass. Over supper that night, the lodge owner came to apologize for the incident: "I am so sorry … We've asked [the neighbouring conservancy] to keep the cows away from this stretch of the river for the guests' sake but I guess I need to remind them," she explained. "I'll definitely be calling somebody about that. I hope it didn't ruin your tea." Clearly, the right for pastoralists and pastoralism to be present in conservancy landscapes – to disrupt, even for a fleeting moment, imaginaries of Africa as pristine and wild – is something heavily mediated and controlled by settler ecologists.

Expanding Settler Ecologies into Pastoral Rangelands

The Community Conservancy Movement

It is not only private conservancies that have undergone repeopling in recent years. Settler ecologists have also taken on an active role in establishing new types of conservation spaces on pastoralists' communal rangelands. These new spaces are community conservancies – communal landholdings where pastoralism and conservation are encouraged to coexist and complement each other. The ecological relations that exist within communal rangelands are a product of centuries of management under pastoralist systems. However, the creation of new community conservancies is meant to reconfigure these landscapes so that they better reflect the ideals of the global conservation community. Many community conservancies are being made in the image of private conservancies – spaces that exist for biodiversity conservation but also include pastoralists in select roles, as long as they follow certain rules.

Much like LTM initiatives, the NRT has spearheaded the community conservancy movement in this region. Even though the NRT has described itself as an Indigenous organization (NRT 2013, 8), the organization itself and the wider community conservancy movement in this part of Kenya can be traced back to the early initiative of settlers. In the 1970s, Ian Craig, whose family owned Lewa, was on safari in Matthews Range (Lenkiyio Hills) when he encountered a family of elephants that had been slaughtered. Our acquaintance in the Suzuki introduced earlier in this chapter, who we call Daniel, recounts the narrative in the following way:

> The streams around there [where the elephants were slaughtered] were actually flowing with blood. You can imagine how many litres of blood an elephant has. Imagine ten down in one spot. That became the turning point for Mr. Ian Craig, because he just looked at a whole family of elephants down and he definitely knew that if all of us continued this way, in the next few years, we would have nothing left for the coming generations.

As detailed in the previous chapter, over the coming years, the Craigs would begin to convert their own land for wildlife preservation following this life-changing encounter. However, they soon noticed an underlying problem with the private conservancy model:

> [Certain] species, such as elephant, giraffe, zebra and lion, could move freely across the landscape. Frequently, animals that had spent time at Lewa were butchered for meat, or in the case of elephants killed for their ivory. It became clear that Lewa's wildlife would only flourish with the help of surrounding communities. (NRT 2013, 6)

In response to this problem, which was not only spatial but social and ecological, Lewa is said to have adopted a new mantra: "How do you save a species? You save a community" (Render Loyalty 2018). From

Craig's perspective, Lewa's conservation efforts risked being undermined by circumstances beyond the conservancy's property lines, unless pastoralists and their livestock were enrolled in a shared vision for conservation across the landscape.

The first stop in Craig's personal mission was the creation of a community conservancy on Il Ng'wesi land, which directly borders Lewa. Craig had good relationships with leaders in this area, including Simon ole Kinyaga, who was a childhood friend (Thomas 2000). Craig set out to convince Il Ng'wesi to develop their own conservancy, complete with eco-tourism ventures such as those at Lewa, that would attract international guests. The rationale was that attaching material benefits to wildlife would give this pastoral community incentive to tolerate and protect wildlife on their land. Initially, Il Ng'wesi was sceptical about the idea. Many thought, "This is just another *mzungu* [white person] trying to steal our land," explained our acquaintance, Daniel. They feared "their land would be turned into a national park or wildlife sanctuary, and cattle would be excluded ... However, Ian kept coming back" (NRT 2013, 7).

In response to their fears, Craig invested significant time and resources into winning people over, even taking village elders on excursions to Lewa Wildlife Conservancy and other conservancies in the Maasai Mara to show them the benefits of wildlife conservation and safari tourism. According to the NRT, "It took time, but eventually the community accepted that the idea of setting up a conservancy was noble and good" (NRT 2013, 7). As Daniel put it, "Ian was very, very determined." In 1995, Il Ng'wesi voted in favour of establishing a conservancy in the group ranch, setting aside 8,675 ha for the initiative, which at the time was over half of the community's land. Craig also raised money from Europe and the United States, along with KWS, to build an eco-lodge on a hill in Il Ng'wesi with a clear view of the landscape all the way to the sacred mountain of Ololokwe in Samburu.

Once Il Ng'wesi became operational, Craig approached other communities to the north of Lewa about forming conservancies on their land. The next to agree was a Samburu community, Namunyak. Like Il Ng'wesi, Namunyak also entered a partnership with Lewa Wildlife Conservancy in 1995 and has since become home to Reteti Elephant Sanctuary, which is discussed in chapter 4. Next came Lekurruki, which borders Il Ng'wesi to the north, with an eco-lodge called Tassia that was built in the 1990s with support from Lewa and Borana and has been externally owned and operated by settlers. By the early 2000s, more and more communities were following suit, forming collaborative relationships with Lewa to protect wildlife and setting aside parts of their land for conservancies and eco-tourism ventures.

The NRT was eventually established in 2004 to coordinate and support the growing number of community conservancies entering collaborative arrangements with Lewa Wildlife Conservancy. Nearly twenty years later, the NRT oversees a total of forty-three community conservancies encompassing over 6.2 million ha in ten different counties in Kenya (NRT 2021), and recently expanding into Uganda. Over time, the organization's role has evolved from simply establishing and managing conservancies to "raising awareness, setting standards, overseeing board elections and training board members … rais[ing] funds for the conservancies and provid[ing] advice on how to manage their affairs" (2015a, 8). The NRT also provides training, advises on security and grassland management, and helps to broker agreements between communities and conservation and tourism investors. Finally, the NRT monitors the performance of each conservancy, providing donors, investors, and government funders with a veil of oversight and quality assurance for their investment.

Although the community conservancy movement in Kenya is bigger than the NRT alone, this umbrella organization has played a vital role in advancing Kenya's community conservancy movement. Unlike other spaces for wildlife conservation, the NRT emphasizes the integration of existing land uses and livelihoods into conservancies. According to the NRT, community conservancies do not exclude or restrict pastoralists or livestock from conservancy areas once they are established. Rather, the NRT model is said to develop conservancies around the needs of pastoralists and the ecological benefits of pastoralism. For this reason, NRT conservancies are often framed as an alternative conservation model in Laikipia where it is common to see people, livestock, and wildlife sharing land and resources.

Over time, the unique mix of people, livestock, and wildlife on these lands has become a key selling point for tourism ventures in NRT conservancies: tourists are sold on the promise of an experience that involves encounters with wildlife and local cultures, including features in the landscape that are of cultural significance to pastoral communities. Tourists come to NRT conservancies to stay in eco-lodges that are constructed in a unique architectural style that combines the colonial safari aesthetic with elements of local material culture. Infinity pools may be surrounded by a *boma*-type fence, with support beams and furniture having been fashioned from trees felled in the surrounding landscape – usually by elephants or other natural forces, it is said. The hides of cattle and goats may be spread across floors alongside those of zebras and impalas, and walls may be adorned with black-and-white photos of herders in traditional dress, important ceremonies, or camel caravans, all meant to illustrate what previously existed on the sites that eco-lodges now occupy. Although eco-lodges on NRT properties are infused with a more rustic vibe than many other lodges in the region, there is no loss of luxury, with these lodges still catering to super elites, including members of the British Royal Family.

While visiting such eco-lodges, tourists venture out into the wider conservancies with the hope of encountering endemic species that thrive in dry lowland environments, viewing Beisa oryxes, gerenuks, Grévy's zebras, reticulated giraffes, and Somali ostriches alongside cattle, all in the same outing. Tourists are likely to be guided by individuals who were born and raised in nearby settlements and herded cattle as young men on what is now conservancy land. Their guides will point out plants, like the whistling thorn acacia, explaining that these spiny trees do not simply have a unique symbiotic relationship with multiple ant species, but are also used in the construction of *boma*. Guides might break a *Salvadora persica* branch, demonstrating its use as a toothbrush to guests and encouraging them to follow suit, explaining that giraffes enjoy consuming the leaves of this tree, which contains a high ratio of fluoride (see photo 3.2). Tourists may also be taken to nearby *boma* and homesteads to interact with livestock and people in the wider landscape. Visiting a *boma* at dusk as cattle return from grazing is an exciting event for many tourists, one made all the more exhilarating when lions can be heard calling in the distance as darkness sets in.

These types of cultural experiences and encounters are valued by many tourists, including those who have been sold on the idea that their holiday contributes to conserving the environment and supporting local populations. After spending time in a NRT conservancy, one such guest wrote the following:

> This place is pure magic ... For me, the greatest interest lay in learning how the community decided to embrace the concepts of conserving wild life and holistic range management on this fragile semi-arid land with their traditional "wealth" being measured in flocks ... And in seeing how proud they are of what they are achieving! Now they see and understand that wild animals can bring them income in the form of tourists ... And the animals are returning ... I also really enjoyed music played on a home made Masaai guitar. The people are the greatest attraction for me, though I saw kudu, elephant, wild dog, elands and many other animals both at the water hole and on the game drives. (Anonymous Reviewer 2013)

As this quote illustrates, part of the allure and novelty of visiting a NRT conservancy is the perceived authenticity of the experience. At the same time, this quote reflects how colonial imaginaries continue to inform tourists' stereotypes of Africa's wildlife as being in need of saving from

Photo 3.2. A guide holding a piece of *Salvadora persica*, demonstrating its use as a toothbrush to guests

local populations, who are perceived as only just beginning to understand and value wildlife.

Reconfiguring Animal–Human Relations on Rangelands

Although the benefits of incorporating pastoralists and their livestock into settler ecologies are quite clear, the implications for pastoralists are more complicated. Of course, there are economic opportunities created by an eco-tourism sector that relies on local labour. Pastoralists also benefit from limited access to private conservancy land for grazing and improvements to community conservancy land through holistic rangeland management. Yet, alongside these opportunities, repeopling has brought about a reconfiguration of animal–human relations on conservancy land that also creates new risks for pastoralists and their livestock.

Pastoralists living in NRT conservancies explain that, as a result of the success of the NRT model, wildlife numbers are increasing. As one elder explained during a conversation we had in Il Ng'wesi in 2019,

> What has changed in recent years is that these wild animals have become so many. And what has made these changes? What has brought the buffalo, lion, rhino, and elephant to be too much disturbing our people so much? It is these lodges. Those with lodges and land are the ones who have brought all of this wildlife.

This knowledge is corroborated in documents produced by the NRT and KWS. For example, as land use surrounding Il Ng'wesi Conservancy has changed from primarily livestock keeping to include biodiversity conservation and eco-tourism, wildlife populations have flourished across the entire region. This larger landscape now contains 46 per cent of Kenya's black rhinos, 90 per cent of the global population of Grévy's zebras, and over 7,000 elephants (LWC 2017).

Alongside growing wildlife populations, wildlife behaviour is also changing. Earlier in this chapter, we explained that intimate encounters between tourists and wildlife are essential to settler ecologies. Although most wild animals are not naturally disposed to sharing close quarters with people, private conservancies in Laikipia have gone to great lengths to curate ecosystems and landscapes in ways that enable tourists to experience close-up and intimate encounters with rare and highly endangered species. This has been nurtured among individual animals and populations over the course of multiple years or even generations, with the assistance of strategically placed dams, salt licks, and watering holes in spaces frequented by tourists.

Some pastoralists suggest that these practices are desensitizing wildlife to human activity and conditioning wildlife to spend more time in closer proximity to people. Past methods used by Il Ng'wesi communities to deter elephants from human settlements no longer work, as elephants have come to associate human settlements with food and water. One elder in Il Ng'wesi explains:

> Back when I was young, livestock would lick the mineral salt while out in the grazing fields. The salt was never brought home ... Now, tourism lodges are introducing wildlife to salt ... Lots of mineral salt licks are left out in the open for them, to attract them close by. Both domestic and wild animals [in conservancies] get to lick the salt, but the elephants like it the most. Eventually, elephants get to know where they can find salt. If they see humans, they expect to also find salt ... They proceed to the houses where the salt lick is stored. They insert their trunks into the house, in a bid to lick the mineral salt. We just wait to see what happens.

A second elder adds the following:

> In the past, when [elephants] smelled smoke they went away, but it's hard now to prevent elephant conflicts, as you could light twenty fires and the elephants would still come. The elephants are coming to the *bomas* for salts, water, and food. Our elders have not seen this behaviour before. [Elephants] used to fear the smoke.

These elders' ideas about underlying factors contributing to changes in elephant behaviour point to practices and strategies used to attract wildlife to conservancies for the sake of tourists. From the perspective of these elder pastoralists, the rise of conservancies with eco-lodges is leading wild species to actively seek out spaces where humans live and go about their daily routines – as they are now accustomed to finding water, salt, or food in such spaces.

All too often, the intimate encounters so desired by tourists turn into lethal entanglements in landscapes where people and livestock move about on foot, and where water and food resources are shared. Elephants in particular have become a constant source of risk to pastoralists. During our research, we were taken to multiple sites of recent conflict between elephants and people. At many of these sites, a human life had been lost. As different elders led us to these sites – where they had lost a brother, a sister, or a son – they gave play-by-play accounts of the events that unfolded there. "The two men were walking here, away from the shops. They came around this bend and found five elephants." As this story was told, we walked in the footsteps of the two men being described. "One elephant saw them and charged immediately. The two men ran. One ran down there [behind a bush]. The other fell down here. He was killed instantly" (see photo 3.3). At another site we visited, a man attempted to escape from an elephant by climbing an ant hill, but he was overtaken before he could climb high enough. He died in a hospital a few days later.

What is interesting about all the sites we visited is that they are not located deep inside the core areas of conservancies. They are in settled areas, near homes, schools, and shops. One man was killed on a murram road frequently used by motorists and pedestrians. Another was killed on a dirt road leading to Nadungoro Plateau, where livestock are grazed. Yet another elder we spent time with described how he was pinned to the ground by an elephant in front of his house. He showed us the place where the elephant's two tusks passed on either side of his torso and pierced the rocky soil.

These events mirror research conducted on human–wildlife interactions in other parts of eastern and southern Africa, which has found that permanent artificial food and water sources can "strongly modify" elephant movement and can alter elephant behaviours (Loarie, Van Aarde, and Pimm 2009). When such spaces are created, wildlife may choose to stay in close proximity to humans – even if it leads them to feel more anxious and stressed – if the perceived benefits from interacting with humans are high (Maréchal et al. 2016; Szott, Pretorius, and Koyama 2019). Ultimately, as conservation landscapes have been repeopled, relationships between pastoralists, livestock, and wildlife are being reconfigured in ways that complicate coexistence.

Photo 3.3. Site where a man was killed by an elephant along Borana-Sanga Road

Making Settler Ecologies: Repeopling

"I know that safari guide, there. I worked with him before," Daniel whispers as we peer out the front window of the military-green Toyota Land Cruiser he is driving. We are parked in the heat of the midday sun, sipping on bottles of warm Coca-Cola and snacking on Manji biscuits, observing a small pride of lions resting in the shade of a tree just a few metres ahead. Less than thirty minutes before spotting the lions, Daniel collected us from an airstrip in Lewa Wildlife Conservancy, where it took our pilot two attempts to land due to a herd of zebras on the strip.

We are driving north across Lewa towards Il Ng'wesi Eco-Lodge, where we will spend the next six days in conversation with conservancy staff and residents as part of the early research for this book. We plan to observe first-hand from a tourist perspective how safaris at Il Ng'wesi compare to safaris in settler conservancies and ranches. For this reason, we decided to embrace the full safari experience by not driving our usual route to Nanyuki or Isiolo, instead departing from Nairobi's Wilson Airport in a packed Safarilink Cessna that seemed in no particular hurry to arrive at its destination.

This is the first time we meet Daniel, who will become essential to our research for the next year or two. After touching down in Lewa, we explain the purpose of our trip to him, which may be why he chooses to direct our attention to the nearby safari guide when we stop to watch the lions. We turn to look in the direction to which Daniel is motioning and see a white safari guide sitting in the driver's seat of a nearby vehicle – also a Land Cruiser painted the same shade of green. The Land Cruiser is filled with tourists wearing khaki suits, wide-brimmed hats, and dark sunglasses, speaking loudly in American accents as they peer through binoculars at the lions panting and shaking away flies from their ears, eyes, and noses. The guide is shushing them: "Lower your voices, lower your voices," he whispers in a gentle yet authoritative tone. Daniel continues, explaining that when there are no clients staying at Il Ng'wesi, he sometimes takes on short-term work with this guide. When he does, Daniel is primarily responsible for tracking game and serving sundowners. "I worked with him recently," Daniel explains, "and he told me that I should not speak to him in English in front of clients. I should only speak Swahili." After a short pause, Daniel continues, "Imagine being told not to speak English. That is worse than colonialism."

This story encompasses one of the key messages we hope to get across in this chapter. Although pastoralists and their livestock are now being incorporated into settler ecologies they were previously excluded from, their inclusion is often dictated and determined by settler ecologists. The story above illustrates an everyday example of this, as Daniel's employment was contingent on him pretending he could not speak English so the *mzungu* (white) guide could perform his notion of white superiority and linguistic belonging in front of tourists (see McIntosh 2016, 2017, on linguistic belonging). To be included in settler ecologies and granted access to some of the benefits accruing from these ecologies, Daniel and many others must agree to the stipulations set out by settler ecologists that problematically demand, publicize, and preserve their Otherness in the landscape.

For pastoralists like Daniel working in the tourism sector, the rules and stipulations sometimes involve dressing and performing the role of the native – silently tracking wildlife for a white guide or dancing to traditional drums, stripped of any other meaning or purpose or form of existence in the eyes of tourists. Meanwhile, the difficult, intellectual work of conservation is framed as the responsibility of settler ecologists who are portrayed as having the knowledge required to educate pastoralists on how best to live with nature and conserve wildlife. This includes white Kenyans, foreign conservationists, and national conservation organizations, including the NRT, seen as having the correct education, experience, and knowledge to make decisions about where and how pastoralists should be included in conservation. Like in other settler-colonial contexts, Indigenous and racialized people are still less likely to be seen by the international conservation community as holding the "right" type of knowledge for conservation and continue to be placed in roles scripted by powerful actors asserting the authority of their knowledge (Kepe 2009).

If pastoralists diverge from settler ecologists in how they wish to use, manage, or develop the landscape, they risk forgoing the right to exist in the landscape altogether. The neopaternalism is effectively illustrated by the heavy-handed and punitive responses to the invasions of settler conservancies and ranches discussed at the beginning of this chapter, where pastoralists returning to graze on their ancestral territories were forcefully removed from ranches and their livestock slaughtered in the process. This paternalism can also be seen at work in the microaggressions directed at pastoralists shrugged off as misbehaving and ignorant, exemplified by the lodge manager who rang up their neighbouring conservancy to complain about the cattle that interrupted our morning tea. Even on pastoralists' own land, settler ecologists sometimes exercise an uncomfortable amount of influence over the terms of pastoralists' presence in the landscape. For example, eco-tourism investors in community conservancies have recently attempted to insert clauses in their contracts that forbid further development by pastoralists on their own land – including eco-tourism or wildlife ventures – even when these investors are only leasing a very small amount of conservancy land.

Understanding why settler ecologists are going to such efforts to fix pastoralists in the landscape and control their herds and herding practices – rather than simply exclude them – requires consideration of what is to be gained by settler ecologists when pastoralists play by their rules. First, ecologically, the positive effects of controlling how pastoralists are brought into conservation landscapes are quite clear. While large populations of livestock are known to negatively influence the quality of rangeland and the distribution and abundance of wildlife species (Kirathe et al. 2021), the planned grazing of smaller herds has long been recognized as a tool for improving rangeland conditions for both wildlife and livestock (Savory and Butterfield 1999; Neely and Hatfield 2007). Proponents of integrating cattle, in particular, into conservation areas argue that strategically placed and concentrated herds can help restore rangeland conditions by breaking up compacted soil, increasing water infiltration and plant growth, improving seed burial through the laying of litter and dunging effects, and enhancing growth of palatable species (Odadi, Fargione, Rubenstein 2017). There is also evidence that allowing for shared use of rangelands by pastoralists and livestock alongside wildlife decreases the likelihood of rangeland fragmentation, as it reduces pressure to convert land for other uses. Maintaining healthy, intact rangelands in turn supports wildlife.

Repeopling conservation landscapes also enhances the moral legitimacy of settler ecologists, helping to secure a future for wildlife in the wider landscape. NRT-T and other LTM initiatives have been influential in this area, mitigating hostility towards existing conservancies in Laikipia and the expansion of community conservancies into surrounding areas. Settler conservancies that participate in LTM initiatives are described as "beacons of sustainability" and "conservation trailblazers of the highest order" (Jones 2019; Marshall 2021). The owners and managers of these conservancies are recognized with prestigious awards for their participation in such schemes, such as Ian Craig who was awarded the Order of the British Empire in 2016 for his role in creating a "world-renowned catalyst and model for conservation that protects endangered species and promotes the development of neighbouring communities" (Capital FM 2016, para. 2). Ol Pejeta Conservancy was also given a World Travel Market Responsible Tourism Award for being a global leader in balancing benefits for local communities and wildlife conservation. Clearly, incorporating pastoralists and their livestock into conservation spaces has benefited the local and global reputation of Laikipia's conservancies.

Finally, as settler ecologists have invited pastoralists and their livestock into conservation landscapes, they have become eligible for funds and grants that would not be available to them otherwise. The United States Agency for International Development (USAID), UK Aid, and other European and UN agencies also support Ol Pejeta Conservancy, while Lewa Wildlife Conservancy has been a recipient of multiple substantive grants from USAID and Global Environmental Facility for its community livelihood initiatives. The average settler conservancy or ranch would not qualify for the grants and financing offered by these aid and development agencies. It is only through repeopling – by incorporating traditional development subjects, initiatives, and programs into conservation and eco-tourism landscapes – that these revenue streams are opened up. In these ways combined, many conservancies and conservation organizations have come to play an essential role in maintaining settler ecologies in Laikipia and across the wider region by spearheading repeopling initiatives that shore up support and free up new funds for conservancies.

4 Rescuing

During the evening of 18 September 2006, Sheldrick Wildlife Trust received a report from Ol Pejeta Conservancy in Laikipia about an elephant cow that had died and whose young calf was standing helplessly by her body. Batian Craig, manager of Ol Pejeta at the time, assigned a guard to keep watch over the calf and protect it overnight from other animals. The next day, the rescue team from Sheldrick Wildlife Trust arrived at Ol Pejeta to rescue the calf, capturing her and transferring her by vehicle and plane to their nursery for orphaned elephants in the Langata suburb of Nairobi. Batian Craig and the rescue team explained that the calf's mother had come to the ranch because she was unwell, "knowing that death was inevitable, and her calf would be safer [at Ol Pejeta] where elephants and others are welcome and protected" (SWT, n.d.-d). The calf was named Lenana, as she was found near Mount Kenya. Mount Kenya's three highest peaks – Batian, Nelion, and Lenana – are named after Maasai leaders who negotiated with the British during the early years of colonization.

For the next two years, Lenana was cared for at the nursery in Langata, on the edge of Nairobi National Park, alongside other orphaned elephants. She was fed milk by bottle every three hours, provided with toys and games each day, and slept alongside her carers each night. In 2008, it was decided that Lenana was mature and healthy enough to be moved to the Sheldrick Wildlife Trust's Ithumba Reintegration Unit in Tsavo East National Park, where orphaned elephants are slowly transitioned to independent life in a natural environment while still living in a fully protected area patrolled by Sheldrick Wildlife Trust and KWS rangers. The Trust reports that Lenana successfully "went wild," and continues to roam freely to this day in a herd of other ex-orphans (SWT, n.d.-b). In August 2020, workers at the integration unit spotted a tiny calf with Lenana. The birth of Lenana's firstborn calf was a moment of pride for Sheldrick Wildlife Trust: not only had Lenana been saved after the death of her mother, but she was now contributing to rescuing the species as a whole.

According to the International Union for the Conservation of Nature (IUCN), the world has entered another mass extinction event. Recent headlines informed by the Intergovernmental Science-Policy Platform on Biodiversity and Ecosystem Services (IPBES) Global Assessment Report on Biodiversity and Ecosystem Services in 2019 warn that one million animal and plant species may face extinction in the coming decades (Díaz et al. 2019). The focus on fighting against extinction in biodiversity conservation has become an effective and affective way for policymakers, NGOs, and the media to signal the urgency of the bio-diversity crisis. As Turnhout and Purvis argue, "[r]epresenting extinc-tion, the loss of species from specific ecosystems or the entire planet, is a powerful way to express biodiversity loss," as it can lead to strong psychological responses (2020, 673). This may be especially true where the extinction of iconic flagship species is concerned.

Conservation organizations have worked hard to develop a sense of connection between the general public and certain species. For decades now, parents in Canada, the UK, and other relatively wealthy societies have bought stuffed toys of endangered animals from environmental charities, such as the World Wide Fund for Nature (WWF), for their chil-dren to cuddle. It is easy to accept that bearing witness to the extinction of real-life elephants and rhinos, pandas and polar bears, and orang-utans and tigers would elicit emotions of mortal dread, fear, and terror for these children and parents. As Garlick and Symons (2020) explain, heightened anxiety about species loss is forging new geographies of extinction where highly affective (i.e., emotional) conservation activi-ties, policies, and programs help nurture intimate connections between people and species on the brink of extinction.

Due to these new geographies of extinction, wildlife sanctuaries, like the Sheldrick Wildlife Trust Centre in Nairobi where Lenana was raised, have come to play an increasingly important role in Kenya's conserva-tion landscape. Sanctuaries all across the country have emerged as a driving force behind conservation action and finance. We define sanctu-aries as spaces where care is offered to animals that are ailing, orphaned, or have been rescued from the wild or confiscated from unlawful cap-tivity. In Laikipia and surrounding areas, several sanctuaries provide care to abandoned and orphaned elephants, trafficked chimpanzees, and injured or endangered animals that require strict surveillance, such as northern white rhinos. The stated mission of these enterprises is often to rescue, raise, and rehabilitate animals so that they can contribute to growing the population of their species and, in some cases, bring their species back from the precipice of extinction.

In many cases, the end goal of rescuing and providing sanctuary to orphaned, injured, and confiscated animals is to reintroduce them to the wild – which tends to be understood quite loosely as any patch of land resembling natural habitat, including that within the confines of fenced private conservancies – where it is hoped they will reproduce and grow local populations of their species. When it is decided that an animal is not fit to be released into the wild for whatever reason, the animal may still be enrolled in the work of saving its species or other species. For example, rescued animals may be used to produce knowledge for conservation through research or put to work in captive-breeding and stocking programs. In some cases, unreleasable animals may also be turned into ambassadors for their species, carrying out affective labour through contact and interaction with humans that generate feelings of intimacy, excitement, sympathy, and wonder (Bersaglio and Margulies 2022). More conservancies in Kenya are relying on these types of encounters with ailing, helpless, and endangered animals as a source of revenue.

In this chapter, we reflect on how wildlife sanctuaries impact ecologies and alter human–nonhuman relations in wider ecosystems and landscapes. More specifically, we aim to show how sanctuaries contribute to creating quasi-wild animals that evoke and, in some cases, require ongoing interventions from settler ecologists, thereby legitimizing settlers' continued presence in contested landscapes. In addition to providing another example of how animals can serve as conduits for settler colonialism, this chapter also responds to the fact that few have critically engaged with the implications of sanctuaries for wider ecologies (for recent exceptions, see Parreñas 2012, 2016, 2018). Instead, national parks, community conservation areas, and zoos have attracted the bulk of critical scholarly attention (for examples, see Dressler et al. 2010; Shukin 2011; Adams and Mulligan 2012; Brockington, Duffy, and Igoe 2012; Büscher et al. 2012; Bigger et al. 2018; Neves 2019).

This chapter begins by briefly historicizing the rise of what we call the conservation-by-sanctuary movement in Kenya. Here, we introduce the settlers who set in motion this conservation movement by establishing sanctuaries and organizations to rehabilitate wildlife in the years following Kenya's independence. The next part of the chapter consists of three sections that describe three different approaches of the conservation-by-sanctuary movement. These are rescuing to release, rescuing to relocate (and relocating to rescue), and rescuing to retain. In the final section of this chapter, we reflect on how these different modes of rescuing work together to produce relations between settler peoples, animals, and wider landscapes that serve to sustain structures of settler colonialism.

The Rise of Rescuers

Although the sanctuary movement has only recently become a significant trend in Kenya, the rescuing and refuging of wild animals by settlers has a long and somewhat contradictory history. As discussed in earlier chapters, during the early days of colonial settlement, settlers were primarily interested in eliminating wildlife on their properties and hunting them for sport. Over the years, settler entanglements with wildlife grew less lethal, as many desired to be surrounded by living wild animals – rather than trophies suspended on corridor walls and above fireplaces. As settlers became concerned with rewilding their properties and wider landscapes around the twentieth century, it became increasingly common for them to rescue and provide sanctuary to orphaned animals and to release these animals, once grown, into habitat around their homesteads.

There are many biographical accounts of settlers rescuing animals in the early years of settlement (for example, see Huxley 1959). However, Joy Adamson is often credited with the birth of what we call the conservation-by-sanctuary movement in Kenya. Born in 1910 to a wealthy family in Austria-Hungary (now the Czech Republic), Adamson met her third husband, George Adamson, while on safari in Kenya in the 1940s and they soon married. George was a British game warden in Kenya's Northern Frontier District, and in 1956, he shot a lioness who charged one of his colleagues only to find three young cubs of the lioness nearby. George brought the cubs home to Joy, where the smallest, Elsa, was raised. The other two cubs were shipped off to Rotterdam Zoo. As Elsa grew, the couple decided they should release her, but first they had to teach her the skills she would need to survive in the wild. Elsa's release was a success and Elsa thrived in the wild, soon having her own cubs.

Over the coming years, the Adamsons attempted to replicate and improve their model for rescuing, rehabilitating, and releasing lions. However, in the 1970s, lions being cared for by the Adamsons mauled carers and visitors on multiple occasions, tarnishing the reputation of their efforts. In the 1980s, around the time of Joy's death, the Government of Kenya placed restrictions on their operations, refusing to allow any new cubs to enter their program. Eventually the program would be reinstated and George would go on to rescue and release several more cubs. The Adamsons' contributions to conservation, seen as innovative at the time, made them widely acclaimed and became a model replicated across Kenya, other parts of Africa, and around the world. Many will likely be familiar with the story of Elsa and the Adamsons, as Joy's book, *Born Free*, became a bestseller and inspired a Hollywood film in 1966. Virginia McKenna and Will Travers, who played Joy and George Adamson in the film, also established an international animal welfare charity called Born Free, which still operates today.

Sheldrick Wildlife Trust has a similar history. Long before the Trust was established, Daphne Sheldrick, wife of David Sheldrick, a national park warden in Kenya, was rescuing and raising wild animals. In her autobiography, Daphne Sheldrick recounts growing up with an eclectic assortment of creatures in her family's garden, including Bunty the impala, Jimmy the kudu, and Baby the eland. However, much like Joy, it was not until her own family shot and killed an animal with babies that Daphne Sheldrick's dedication to raising orphaned animals began. As the autobiography explains, in rather grim terms,

> I learned a great deal, not least from an orphan of the time named Punda, a tiny zebra foal. Out one day on the plains, a heavily pregnant mare fell to Dario's gun, and when one of the helpers immediately opened up her stomach cavity to haul out the viscera there was a uterus in which a fetus was stirring, obviously waiting to be born. A quick slash of the knife opened the bag and the baby, all wet and sticky and kicking feebly, gasped to draw its first breath ... When my father arrived on the scene he extracted some of the vital colostrum milk from the mother's still warm udder to reinforce the foal's natural immune system ... Having staggered on to wobbly legs, he walked straight up to me with an innocence and an implicit trust that touched my heart and made me want to protect him forever ... I mixed the colostrum milk with sweetened condensed milk in a baby's bottle and fed him with as much love as I could show. (Sheldrick 2012, 124–5)

The experience Daphne Sheldrick gained raising Punda soon motivated her to take in other orphaned animals, including elephants and rhinos.

As previously discussed, a profound shift took place in the way Kenya's settlers understood themselves and identified as a collective throughout the post-independence era. Increasingly, they saw themselves as guardians and protectors of Kenya and its nature. Ironically, much of the nature that they endeavoured to protect and save was endangered, at least in part, as a consequence of the hunting done by settlers. The following excerpt from Daphne Sheldrick captures this shift in how settlers identified and understood their purpose, not just in the colony but on Earth itself. Reflecting on the orphaned animals she raised from infancy, Daphne Sheldrick explains:

> A very special feeling flooded over me when I contemplated my orphans … a feeling of great achievement; a warm glow of pride; of deep satisfaction and contentment; a feeling of identity – almost even of creation, for by being instrumental in giving life to one animal, many others had a chance to live. It was, I suppose, a feeling of worthwhile contribution and success, of having led a life that was constructive rather than destructive for only creativity brings true peace to one's soul. So easy to destroy, so difficult to create, but also so much more fulfilled in every way. One last triumph I longed for, was to be able to look with pride on an animal like an elephant, born of a mother I had nurtured from the start, and think, deep inside my heart, "But for me …" (Sheldrick as quoted in SWT 2019, para. 3)

The deep sense of attachment and pride Daphne Sheldrick felt towards the animals for which she provided and oversaw care was picked up by Western media and conservation organizations, which named Daphne Sheldrick "Mama elephant" in the write-ups and obituaries that followed the announcement of her death in 2018 (Laffrey 2018).

Over the coming years, the idea that settlers ought to remain in Kenya to rescue and protect animals would become increasingly pervasive, especially following Kenya's independence. Kenya's white settler community, along with conservationists from other parts of the world, "expressed pessimism that conservation would not be given the priority it deserved by the new African government" (Waithaka 2012, 28). This led some settlers to double down on their view of themselves as nature's last bastion of hope. For example, as Italian-born Kuki Gallman, who settled in Laikipia in the late 1980s, writes in her famous memoir, *I Dreamed of Africa*,

> Now, that Ol Ari Nyiro [the property acquired by Gallman] has become an island of life surrounded by parched landscapes, to reward the dedication of my lifetime … I watch calmly the endless savannah I have nurtured, all covered in African olives, acacias and croton … I can spot sometimes a solitary giraffe ambling unhurried from the thickness of the trees, a small family of elephants feeding from the wide branches of the acacia, and I know, with a glory in my heart, that the tears have not been in vain, the days of tireless work and the sleepless nights of watching over the land, ears strained to detect the sound of gunshots, dreading to receive another emergency call, were all worth it, and this is my reward. The wilderness may have shrunk the world over, but not here. I have been her guardian. (Gallman 1991)

This idea of guardianship – as an obliged responsibility to care for something deemed unfit or unworthy to care for itself and defend its own best interest – reflects a paternalism that has defined governance in settler societies for centuries (Oloruntoba 2020).

Interestingly, as our discussion of rescuing so far implies, the (neo) paternalism pervasive in the rise of conservation-by-sanctuary in Kenya is heavily gendered and largely enacted by individuals that society would see as maternal rather than paternal. While men in settler families filled outward-facing positions in the colonial administration – including roles as game wardens and park rangers – settler women spent far more time contributing to reproductive labour, overseeing women labourers working in and around homesteads and raising and caring for children and domestic animals. For this reason, it is not altogether surprising that, when men brought home animals orphaned by their guns or found injured while on patrol, women would step up to care for these animals. The gendered nature of rescuing can be seen in other parts of the world as well: women tend to be the primary carers of animals in households (Kruse 1999); most volunteers at animal shelters and rescue groups are women (Rattan, Eagles, and Mair 2012); and many animal rights activists are also women (Gaarder 2011). Literature on the subject suggests that women are more concerned with animal welfare, as it is easier for them to identify and forge solidarities with animals through shared experiences of oppression and subordination (Taylor et al. 2020; see also Donovan 2006).[1]

The Conservation-by-Sanctuary Movement

The knowledge and models initially developed by settlers to rescue orphaned and injured animals continue to be used today across Kenya. As these models have been refined and expanded over time, Kenya's conservation-by-sanctuary movement has become well versed in using the same intimate encounters with iconic flagship species that early settlers wrote about to generate revenues and support for the maintenance of settler ecologies. In this section, we introduce three specific modes used within the conservation-by-sanctuary movement to generate this support.

---

1 Of course, there were exceptions to this rule. For example, Huxley writes about her own first big game hunt, where she stalks and shoots buffalos just like any man would do while on safari. Beryl Markham is another exception, regarded as a rightful adventurer for her skills as a pilot. Yet even "exceptional" women such as Elspeth Huxley, Karen Blixen (Isak Dinesen), and Kuki Gallman had to write themselves into history.

Rescuing to Release

The first mode used in the conservation-by-sanctuary movement involves rescuing orphaned, sick, or injured wild animals with the intention of rehabilitating and releasing them back into the wild. Today, this approach to conservation in Kenya is clearly exemplified by Sheldrick Wildlife Trust, which began rescuing and rehabilitating orphaned elephants more than fifty years ago. The Trust has since expanded its operations to include aerial surveillance, anti-poaching initiatives, eco-lodges, veterinary services, and community projects across Kenya, including in highland areas, such as around Mount Kenya, and lowland areas, such as Amboseli and Tsavo. The Sheldrick Wildlife Trust is among the most famous elephant protection programs in Africa and serves as a model that has been replicated across the continent and in other parts of the world. As of 2018, the Trust had rescued and raised 244 orphaned elephants, with 150 of these being released into the wild (SWT-USA 2018). Sheldrick Wildlife Trust currently has over eighty orphaned elephants in its care across its nursery in Nairobi and three reintegration units, located in Tsavo East National Park and Chyulu Hills National Park.

According to Sheldrick Wildlife Trust, its founder, Dr. Dame Daphne Sheldrick, revolutionized the care of elephants and other orphaned mammals in Kenya by developing a special milk formula. Prior to the development of this formula, few elephants were able to survive the loss of their mother, as elephants are entirely dependent on milk until they are about two years old and continue to rely on milk until the age of four. As such, it is imperative for baby elephants to be fed a formula that can sustain their large bodies and keep them healthy. In addition to developing a formula for infant elephants, Sheldrick also recognized that the composition of an elephant cow's milk changed as its baby grew – and found ways to replicate these changes in her own formula. Over the years, this magic formula would be adapted for other species, such as rhinos, allowing the Trust to raise other species too.

In 1978, Sheldrick was granted permission by the Government of Kenya to live in Nairobi National Park, where she established the Nairobi Elephant Nursery. There, she worked with a team of keepers and a veterinarian to raise orphaned elephants, beginning with just two but expanding the operations over time. Today, the Sheldrick Elephant Nursery remains in operation and is a top tourist destination in Nairobi. Orphaned elephants are fed bottles of milk in the morning and afternoon by keepers in front of crowds of tourists and later led to the bush to browse before being taken to a mud hole in which to bathe and play. In the evening, the elephant calves return to their stalls for a nighttime feed before falling asleep with a keeper, "though never with the same keeper two nights running in case either orphan or keeper becomes too attached to the other" (SWT, n.d.-c). In 1997, the first orphaned black rhinos were brought to the sanctuary to be hand-raised. Later, other types of wildlife arrived at the nursery too, such as giraffes and kudus.

Since its inception, Sheldrick Wildlife Trust has also established multiple reintegration units for orphaned elephants in Tsavo East National Park and Chyulu Hills National Park. Elephants are sent to these locations from the Nairobi Nursery Unit regardless of where in the country they were rescued from. They arrive at the reintegration units still dependent on their keepers for milk and protection during the night, which is provided to them from a stockade, but the elephants spend the day roaming in the bush. At each reintegration unit, Sheldrick Wildlife Trust maintains a water trough and mud bath area, attracting wild herds and providing the orphans with plenty of opportunity for socialization. Over time, orphans are weaned off milk and, as they begin to graze and browse more, they naturally begin to spend longer times away from the stockade. Some elephants will choose to join wild herds while others will form their own herds with other orphaned elephants, becoming fully wild and having wild offspring. Yet, even once reintegrated, ex-orphans regularly return to the stockade for water and to visit resident orphans and their keepers.

In more recent years, the Sheldrick Wildlife Trust has expanded its activities – now focusing on anti-poaching projects supported by aerial surveillance, canine units, and mobile veterinary services – to ensure the safety of elephants after they are released, as well as that of other elephants and species in the parks where reintegration units are located. The anti-poaching teams work in partnership with KWS to provide "eyes in the sky" on elephant populations and "prevent illegal activities and apprehend poaching offenders," not only in Tsavo but in other parts of the country when assistance is needed (SWT, n.d.-a). The Trust operates seven fixed-wing aircraft and two rapid-response helicopters that it uses to surveil the landscape and deploy rangers and its canine unit to the field. The organization also has a team of trained tracker dogs and handlers that detect illegal wildlife products, such as elephant tusks and rhino horns, bushmeat, guns, and ammunition, as well as track suspected poachers.

Finally, Sheldrick Wildlife Trust operates eight mobile veterinary units around the country in collaboration with KWS, with a focus on Mount Kenya. Veterinary units include custom-made vehicles that come fully equipped with darting hatches and dart guns, equipment shelves, vaccine refrigerators, an operating table, and all the necessary medicines and equipment required for rapid and effective veterinary response. Each team is led by a KWS veterinarian and consists of a KWS capture ranger and a driver from Sheldrick Wildlife Trust, prepared to treat a range of species, from elephants and rhinos to giraffes and zebras to lions and leopards. The organization also operates a rapid aerial response medical unit, called the Sky Vet, that rescues and aids wild animals in areas that are isolated or difficult to reach.

Over time, Sheldrick Wildlife Trust has inspired other rescue-to-release programs across the country. In 2016, Reteti Elephant Sanctuary – also known as RESCUE – was established in Namunyuk Conservancy. Namunyak is located in the Samburu lowlands directly north of Mount Kenya, the Borana-Lewa Landscape, and Il Ng'wesi. Reteti takes in orphaned and abandoned elephant calves with an aim to release them back into wild herds surrounding the sanctuary. Although described as the "first community owned elephant orphanage in Africa" (RESCUE, n.d.), the sanctuary itself was co-founded by third-generation settlers in Kenya, Katie Rowe and Jeremy Bastard, who also operate Sarara. Sarara was founded in 1997 by Bastard's parents, Piers and Hilary, with support from Ian Craig.

Mirroring the Sheldrick nursery and reintegration model, orphaned calves spend the day in the wild but return home each night to a stockade where they are fed a bottle before falling asleep with keepers. Also similar to Sheldrick Wildlife Trust, Reteti has expanded the mandate of its operations over time. The sanctuary now participates in rescuing species beyond elephants, including an orphaned gerenuk and giraffe. The sanctuary has also acquired a Super Cub aircraft that it uses for aerial surveillance, explaining that the "vast landscape requires surveillance by air patrols that include keeping track of elephant movements, anti-poaching and post release monitoring of the rewilded orphans" (RESCUE 2020). With access to financial means and military-grade equipment, rescue-to-release initiatives have become incredibly adept at mobilizing "extinction narratives" (Weldemichel 2020) to legitimize interventions in landscapes that go far beyond the one-off rescuing of an orphaned animal.

Alongside rescuing, rehabilitating, and surveillance activities, rescue-to-release initiatives have also become a major draw for tourists who desire intimate and spectacular access to wildlife. For example, tourists can visit Reteti to interact with rescued baby elephants in the stockade before heading out to observe those elephants that are older or that have been released back into their natural habitat in the wider conservancy. Reteti has its own airstrip, so tourists can charter planes from Nairobi for a private visit. Tourists can extend their visit by staying at Sarara, a nearby eco-lodge and luxury tented safari camp in Namunyak Conservancy. Then, when the heat and dust of the Samburu lowlands become overwhelming, tourists retreat to the luxury of Sarara, dipping into a cool pool that, "with its view over the distant mountains, offers the ideal place from which to view the surrounding wilderness in private" (&Beyond, n.d.).

The activities and luxuries available to tourists who visit Reteti and Sarara are packaged as community-owned and community-driven, and are acclaimed for giving back to the community. In a recent documentary about the sanctuary, called *The Guardian Elephant Warriors of Reteti*, this is explained:

> Without the elephants we were all suffering. Our ecosystem was dying. We began to understand that elephants are the heart of this land and its greatest engineers. They guide us to water and bulldoze small trees, keeping the land open and grassy. For us to thrive, the elephants too must thrive. (RESCUE, n.d.)

This narrative has become part of the holiday package that tourists are sold – the idea that visiting Reteti offers more than just a self-indulgent getaway. Tourists are told that visiting Reteti is part of a global fight against the extinction of elephants.

Rescuing to Relocate/Relocating to Rescue

Looking beyond elephants, another mammal that was once nearly eradicated by the economic activities and hunting practices of settlers is the rhino. Although somewhat ironic given settlers' historically antagonistic relationship with the species, it is also not surprising that rhinos have also become a focal point of the sanctuary movement in Kenya. Unlike elephants, though, rhinos are rarely ever rescued to release. Instead, rhinos tend to be rescued and translocated to secure rhino sanctuaries where they live quasi-wild lives under twenty-four-hour surveillance and protection. The underlying premise for rescuing rhinos is also slightly different in comparison to other animals, as they are often rescued from habitats where they are deemed to be at risk and moved to dedicated sanctuaries that function as secure harbours and safe breeding grounds (Knight 2019). As rhino sanctuaries are very expensive to operate, they tend to be found in national parks or private conservancies.[2] In Laikipia, most rhino sanctuaries have been established on former settler ranches,

---

2 By way of example, Lewa Wildlife Conservancy spends over 50 per cent of its annual budget on protecting the 150 rhinos in the conservancy (LWC, n.d.). "Adopting" a rhino in the conservancy for a year costs US$2,500 (LWC, n.d.).

with half of Kenya's entire black rhino population contained on just three private landholdings: Solio, Lewa-Borana, and Ol Pejeta.

Rescuing wildlife that cannot be released is a tricky business. Outside sanctuaries, overpopulation of a specific species is unusual, only occurring on islands or in islands of protected areas. However, enclosed, secure sanctuaries restrict the movement of animals and often end up isolating populations within them from those on the other side of sanctuary fences. Species that are heavily protected and not readily preyed on by carnivores – like rhinos – often reproduce easily in the safety of sanctuaries and large numbers of them within enclosed spaces can be harmful for the environment inside sanctuaries. This in turn risks exposing them to predators, epidemic disease, and starvation if the problem is not addressed. When rhinos take too heavy a toll on their own habitat, the wider landscape may suffer as well, as reduced vegetation caused by high densities of these browsers and grazers can lead to soil erosion and other forms of degradation with implications for biodiversity and ecosystem health. These types of conditions can leave all wildlife within a sanctuary vulnerable to environmental shocks, such as drought or flooding, which can trigger mass die-offs.

Sanctuaries in Laikipia have been grappling with this challenge for over three decades, beginning with Solio Ranch in the 1980s. The impressive breeding performance of Solio's black rhinos meant that the rhinos quickly ate themselves out of food within their enclosed habitat. Solio had a black rhino density of 1.2 per km², where 0.5 per km² would normally be considered high (Patton, Campbell, and Parfet 2007). By early 2007, mainstay food sources for black rhinos, such as *Vachellia drepanolobium* (whistling thorn), had almost disappeared, leading to the depletion of other browsers. Lewa and Borana, which co-manage a shared sanctuary for eighty-seven black rhinos, are now experiencing a similar problem (Patton, Campbell, and Parfet 2007). Black rhinos were first (re)introduced to Lewa in the 1980s with the translocation of fifteen individuals, and the population quickly expanded through additional translocations and natural births (Patton, Campbell, and Parfet 2007). More recently, population growth has stalled due to the high density of rhinos and related depletion of food sources. By 2017, the average annual growth rate of Lewa-Borana's black rhino population reached a low of just 1 per cent while the average annual mortality rate stood at 6.2 per cent (Kimiti, Mwololo, et al. 2017).

Ol Pejeta has dealt with this problem by continually expanding the amount of land available for rhinos, allowing it to now host the largest population of black rhinos in East Africa. Since 1993, Ol Pejeta's black rhino population has grown from 20 to 143 individuals (OPC 2021b). As black rhinos are browsers – unlike white rhinos, which are grazers – this population growth quickly increased pressure on vegetation across the conservancy (Olweny, Wahungu, and Obwoyere 2020). Moreover, as black rhinos are solitary in nature, densely packed populations can lead to behavioural issues (Kimiti, Chege, et al. 2017). For this reason, after being purchased by FFI in 2004, Ol Pejeta increased the area available for rhinos within the conservancy to 36,500 ha by opening up ranch lands that had yet to be used as habitat. While all this was happening, to the south of Ol Pejeta, Solio was also facing a rhino overpopulation problem, not to mention security challenges, as was Ol Jogi to the north. This led to dozens of rhinos being translocated to Ol Pejeta from these sanctuaries (Patton et al. 2010a; 2010b). From a starting population of just four black rhinos in the 1980s, Ol Pejeta is now home to over 130 individuals.

As each of Laikipia's three major rhino sanctuaries have grappled with the environmental challenges that come with containing high densities of rhinos, new sanctuaries have been established and recruited to participate in rescue-to-relocate/relocate-to-rescue initiatives. In 2015, several black rhinos were translocated from Lewa-Borana to Sera Conservancy in Samburu, making Sera the only black rhino sanctuary located in a community conservancy in Kenya. With funding from the NRT, KWS, FFI, Save the Rhino, WWF, the San Diego Zoo, and Tusk Trust, Sera is now acclaimed for the collaborative role it plays in "securing the future of this and other species that remain precariously close to extinction" (TNC 2021). More recently, in 2020, Loisaba Conservancy was given conditional approval by the Government of Kenya to host a new rhino sanctuary. Previously owned by the Italian Count Carletto Ancilotto, Loisaba was recently purchased by TNC and Space for Giants. These organizations raised US$5.1 million for the infrastructure upgrades required to "bring black rhinos back home to Loisaba" (TNC 2021), including rhino-proof fencing, all-terrain vehicles, and other security and surveillance equipment.

Although Sera and Loisaba have recently joined the ranks as approved rhino sanctuaries, only a small handful of properties in Laikipia have been able to meet and retain the stringent and expensive requirements of protecting rhinos on their land. With so few conservancies permitted to have rhinos on their land – an endangered species that carries such moral and symbolic value in the fight against biodiversity loss and extinction – rhino sanctuaries provide their operators with staying power in the landscape. As Bersaglio and Margulies (2022) have argued, the transformation of settler ranches into rhino sanctuaries – with their tight security features, highly armed personnel, and support from powerful conservation actors internationally and within Kenya – has served as a way for settler ecologists to morally justify their role in this particular landscape while also securing their property rights and staving off land redistribution. As one white settler explained of Lewa Wildlife Conservancy, "the

Craigs were smart. They put rhino on their land. That got the whole international community behind them." The same sentiment applies to international conservation organizations, like FFI and TNC, that similarly shore up support for their ongoing presence and influence in Laikipia by rescuing and relocating rhinos on their landholdings.

Rhinos are not the only species being rescued through relocation. Other highly endangered species are also rescued from their natural habitats where they are deemed to be at risk and moved to dedicated sanctuaries. Mountain bongos provide one such example. Mountain bongos are endemic to the more mountainous areas of central Kenya's highlands, such as Mount Kenya and the Aberdare Range, and are classified as critically endangered. This subspecies of bongo has been in decline since the 1890s, when settlers introduced rinderpest-infected cattle to the region.

In the 1960s, Hollywood actor William Holden and TV personality Don Hunt established the Mount Kenya Wildlife Conservancy (previously called Mount Kenya Game Ranch) in an effort to save the mountain bongo. After a hunting and fishing trip to Kenya in 1964, the pair enlisted the help of a former professional hunter and naturalist to establish a sanctuary devoted to the care and rehabilitation of orphaned and injured animals, including mountain bongos. Early on, the conservancy organized the capture of twenty young mountain bongos, which were shipped to zoos in the United States to form a nucleus breeding program. Soon, others became involved in running the sanctuary and supporting bongo rehabilitation, including KWS, Kenya Forestry Service (KFS), the Bongo Surveillance Project (initiated by Mike Prettejohn), Rhino Ark (established by settler Ken Kuhle and later run by past chairperson of KWS Colin Church), the US-based International Bongo Foundation, and the American Association of Zoos and Aquariums.

While mountain bongo populations in Kenya continued to steadily decline into the 2000s due to illegal hunting and agricultural expansion – plummeting from around 500 individuals in the mid-1970s to between fifty and seventy-five in 2010 (KWSa 2019) – the captive breeding population in the United States thrived. In 2004, the American Association of Zoos and Aquariums translocated eighteen zoo-bred bongos to Mount Kenya Wildlife Conservancy to initiate a Kenyan-based breeding program. The bongos were placed in 10 acre breeding enclosures in the foothills of Mount Kenya. This captive breeding population has also grown, with over seventy total births in the past five years (KWS 2019a). In May 2021, it was announced that a further twenty to twenty-five mountain bongos would be moved from the United States to Kenya by December 2021. This translocation was carried out by the newly formed Meru County Bongo and Black Rhino Conservation Trust (MBBR-CT), backed by Lewa Wildlife

Conservancy, Mount Kenya Trust, KWS, KFS, two Community Forest Associations (CFAs), and the Meru County Tourism board.

The twenty-five captive-bred mountain bongos being translocated to Kenya will not be placed in Mount Kenya Wildlife Conservancy. Instead, the bongos will be released into a new 800 acre enclosure, highly secured like those constructed for rhinos, called Mawingu Mountain Bongo Sanctuary (see photo 4.1). Further down the line, additional sanctuaries – or Intensive Protection Zones – will also be established in Ragati, Eburu, Mau, and Aberdares forests for further releases. Each zone will be staffed by a permanent security team of trained rangers, and KWS and KFS will work in collaboration with partners, such as Rhino Ark and Ragati Conservancy, to ensure these mountain bongos are protected within their historic range.

The rescuing of mountain bongos is quite unique as this was reportedly the first species to be repatriated to its original habitat in Kenya with the goal of re-establishing wild populations (UNDP 2003). The act of bringing the bongo back to its ancestral habitat has drawn accolades and recognition from intergovernmental organizations, such as the United Nations Development Programme, as it showcases how a country can go about restoring its wildlife heritage. However, as much as the mountain bongo program has become a project of national significance, it has not evaded the influence of settler ecologists. As mountain bongos are brought back to Kenyan soil, American zoos and scientists and settler-initiated conservation organizations, such as Rhino Ark, remain heavily involved in shaping and securing the future of these species. Moreover, settlers, along with their landholdings, networks, and financing, are also exerting influence over new bongo sanctuaries, with Lewa Wildlife Conservancy playing a central role in the public-private partnership that established these new spaces. As with rhinos, by being involved in the rescuing of such highly endangered and unique species, settler ecologists legitimize their influence over ecological relations in a landscape where it might otherwise be contested.

Rescuing to Retain

As a final example of rescuing through the conservation-by-sanctuary movement, settler ecologists in Kenya are also involved in rescuing animals that are incapable of being released back into the wild – either because they are injured, unwell, or especially vulnerable to hunting or trafficking. The overarching ambition behind rescuing these types of animals is similar to those rescued for release and relocation: the idea is to help grow populations of endangered species or to bring species back

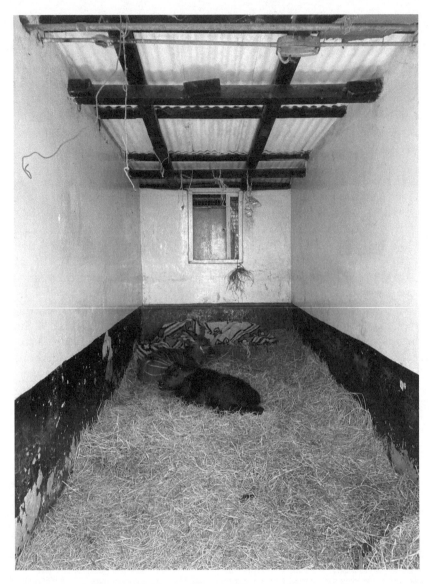

Photo 4.1. A bongo calf failing to thrive in Mawingu Mountain Bongo Sanctuary brought to Mount Kenya Wildlife Conservancy, where it shares a stall with an orphaned buffalo calf

from near extinction. However, by retaining rather than releasing or relo-cating these animals, one key difference in this form of rescuing is the emphasis placed on using rescued animals to attract conservation finance and develop wildlife-specific economies – be these research economies or tourist economies. In this regard, both the practicalities of rescuing to retain and the day-to-day lives of animals retained after being rescued differ from rescuing to release or relocate. We unpack all this further by focusing on one conservancy's efforts to retain critically endangered spe-cies of black rhinos, northern white rhinos, and chimpanzees.

Located just west of Nanyuki Town, Ol Pejeta Conservancy is at the forefront of efforts in Kenya to rescue to retain – providing sanctuary to animals that have been rescued from the wild or wildlife trade but whose injury, trauma, or other circumstances complicate their rehabilita-tion and release. Prior to becoming the fully fledged and internationally financed conservancy it is today, Ol Pejeta was started in the 1980s as a rhino sanctuary called Sweetwaters Game Reserve. Since that time, the conservancy has adopted a conservation-by-sanctuary model that differs from the Sheldrick approach, focusing instead on creating special enclo-sures for injured individuals or small groups of critically endangered species. Within these enclosures, rescued animals are isolated from their free counterparts but still surrounded by natural habitat.

These enclosures are usually attached to platforms from which tourists can observe or interact with the animals. For example, tourists can feed, touch, and take selfies with a black rhino at Ol Pejeta, named Baraka, in one part of the conservancy, and then proceed to observe groups of rescued chimpanzees from a raised platform in another. The enclosures and viewing platforms are usually located within walking distance from small education centres, where tourists are informed about conserving endangered species and can donate to support Ol Pejeta's conservation efforts – for instance, by adopting one of the animals they encountered during their time at the conservancy. Ol Pejeta reports that 80 per cent of its operating costs are covered through tourism with over 70,000 visitors per year, with many of these visitors coming because of the unique and intimate encounters with wildlife that it offers.

Chimpanzees are one of the species that tourists come to visit. Although chimpanzees have the largest range of any great ape, they are not native to Kenya. Their geographical distribution extends from the western tip of Africa to the westernmost regions of Tanzania and Uganda in the Great Lakes region. Nevertheless, when a chimpanzee rescue centre in Burundi was forced to close in 1993 due to civil war, Ol Pejeta provided sanctuary to its chimpanzees. This relocation led to the establishment of Sweetwaters Chimpanzee Sanctuary, a 250 acre

Photo 4.2. Rescued chimpanzee in sanctuary at Ol Pejeta Conservancy

enclosure along the Ewaso Ng'iro River, run by Ol Pejeta Conservancy in collaboration with the Jane Goodall Institute and KWS (see photo 4.2). A member of the Pan African Sanctuary Alliance, Sweetwaters currently has thirty-five chimpanzees and continues to rescue more from illegal trade, underground markets, and other potentially harmful conditions (OPC, n.d.-a). Tourists visiting Ol Pejeta can visit the chimpanzees during feeding times. They can also adopt chimps who steal their hearts or donate to the sanctuary to help ensure "they live out the rest of their days in peace and safety" (OPC, n.d.-b) by funding monthly food provisions, life-saving surgeries, emergency care, and other services. In addition to providing sanctuary and services to chimpanzees, Sweetwaters is also used for research into recovering the chimpanzee species, hosting a range of behavioural and clinical studies each year (Ross and Leinwand 2020).

Just a short drive east of Sweetwaters Chimpanzee Sanctuary is another enclosed sanctuary. Unlike the chimpanzee sanctuary, this enclosure contains an animal that the conservancy allows tourists to interact with from a special constructed platform. Baraka the black rhino originally roamed freely within the boundaries of Ol Pejeta like all the other black rhinos that have been translocated to the reserve; however, after reaching maturity, he was wounded in a territorial fight with another male black rhino and lost sight in his right eye. Baraka later developed a cataract in his left eye, at which point the conservancy moved him into a 100 acre secure enclosure, where he is now visited daily by tourists who take turns feeding, touching, and taking selfies with the blind rhino. Ol Pejeta likens the work done by Baraka to that of an ambassador, explaining that, since being moved to the enclosure, Baraka "is doing a splendid job of being Ol Pejeta's black-rhino ambassador to humans" (OPC, n.d.-e).

Within a stone's throw of Baraka's enclosure is yet another fenced-off area – only this one is more secure and guarded. For over a decade, this 700 acre enclosure has contained a small group of large mammals that are so rare and endangered that they may be considered functionally extinct: northern white rhinos. Ol Pejeta's northern white rhino enclosure now contains a mother-daughter duo named Najin and Fatu. However, it was a male rhino named Sudan, also occupying this enclosure up until his death in 2018, that helped draw the world's attention towards Ol Pejeta's fight against extinction.

In 2009, Sudan, Najin, and Fatu were all translocated from Dvůr Králové Zoo in the Czech Republic to Ol Pejeta Conservancy. The move was justified by the notion that a more natural habitat might help the rhinos breed and reproduce. Not long after the arrival of the trio, Ol Pejeta used Sudan's status as the last male of his kind to raise awareness and finance for its conservation efforts. Like Baraka and other solo rhinos retained in sanctuary by Ol Pejeta in the past, Sudan was framed as an ambassador of his species to humanity. Sudan grabbed international headlines, featured in documentary films and nature shows, and inspired apparel and souvenirs sold by the conservancy. These efforts helped turn Sudan into an international sensation. Tourists from around the world travelled to Laikipia just to see Sudan, including high-profile guests such as Elizabeth Huxley, Khaled Abol Naga, Leonardo DiCaprio, Melissa McCarthy, and Nargis Fakhri. However, Sudan's status as a global celebrity was solidified in 2017 through a collaborative fundraising campaign between Ol Pejeta and Tinder. The campaign involved the creation of a Tinder profile for Sudan, which read, "I'm one of a kind. No seriously, I'm the last male white rhino on the planet earth. I don't mean to be too forward, but the fate of my species literally depends on me. I perform well under pressure ... 6ft tall and 5,000lbs if it matters" (Winter 2017).

The Ol Pejeta–Tinder campaign fell short of raising the targeted US$10 million, bringing in just US$85,000 (OPC, n.d.-d). However, from a public relations perspective, the campaign was highly successful at spreading the news about Ol Pejeta's efforts to begin experimenting with the use of in vitro fertilization (IVF) and other advanced reproductive technologies (ART) on rhinos as a desperate final attempt to pull the species back over the threshold of extinction. Conservation scientists began to prepare for the use of IVF to produce a northern white calf in 2015, when tests confirmed Sudan was incapable of natural reproduction. At that time, veterinarians started to collect samples of Sudan's semen and store them alongside specimens collected from other northern white rhinos before their deaths. After Sudan was euthanized in 2018, efforts to artificially reproduce the species focused mainly on the use of sperm cells procured from these stockpiles of frozen semen. In 2019, five oocytes were collected from both Fatu and her mother, Najin, at Ol Pejeta for use in IVF procedures (Save the Rhino 2020). However, recognizing that IVF was unlikely to succeed, resources were also directed at other forms of ART, including stem cell technology and gene editing (Pilcher 2018).

In 2021, it was announced that scientists from Avantea laboratory in Italy had successfully created nine northern white rhino embryos after extracting eggs produced by Fatu and injecting them with thawed sperm from a deceased northern white rhino bull named Suni. As it seems that neither Fatu nor Najin are capable of carrying a pregnancy to term, scientists plan to transfer the eggs into a southern white rhino surrogate with the hope that this will lead to the birth of a northern white rhino calf. The next stage in this program involves embryo transfer to surrogate southern white females at Ol Pejeta Conservancy. Yet the complex choreography involved in reproducing a species that is considered by many to be functionally extinct is a delicate, intricate affair. There has never been a case in which a viable pregnancy has been realized in a surrogate white rhino (Save the Rhino 2020). Moreover, even if pregnancy is achieved, calves produced using this method may not be genetically diverse enough to sustain a future population of northern white rhinos (Gilliland 2019).

According to Joseph Salazar (2014), we live in a world defined by big science, where Western scientific interests routinely converge with capitalist and other economic interests, national and transnational security interests, public health interests, and population-control measures. In this world, the production of scientific knowledge "depends on access to land and resources for its experiments and observations" (Salazar 2014, 40). It also requires the state's legal support, through permits and land leases, as well as regulatory oversight (Salazar 2014). Due to the legacies of colonization in Laikipia, settler ecologists have access to the land and resources, investments and revenues, government support and approvals, paramilitary actors and technologies, and transnational epistemic communities needed to conduct such research on endangered species, like rhinos. Whether or not this research ultimately leads to success in preserving the northern white rhino, it is choreographed in a way that justifies and creates further demand for Ol Pejeta Conservancy, as well as similar ventures. This suggests that rescuing species should not be seen as a politically neutral force; instead, it is a practice that is shaped by and contributes to the reproduction of existing power relations, with only those that have land, investment, and perceived capacity able to participate in this economy.

Making Settler Ecologies: Rescuing

"Would you like to hold it?" We look at each other and then back at the caretaker standing before us dressed in a green khaki uniform, not sure we heard him correctly. He repeats the question, with a smile: "Would you like to hold it?" Charis replies apprehensively for both of us with a nod and smile in return. "Sure," she says. We follow the caretaker to a small kennel-like enclosure with a concrete floor, chain-link walls, and a wooden house stuffed with hay tucked in the back corner. The caretaker opens the gate, steps in, and begins rummaging around the small wooden house. His hands emerge with a ball of grey fluff. He presses the fluffy creature tight against his chest, walks up to Brock, and gently positions the three-week-old striped hyena in Brock's arms (see photo 4.3).

The baby hyena's fur is much softer than expected, we comment, as we stroke the small, dozing animal. After a moment, the baby hyena repositions itself, letting out a quiet sigh as it gently rests its soft, slightly damp chin in the crook of Brock's arm. Apparently now more comfortable, it closes its eyes and falls asleep, breathing deeply and relaxing its little body. A look of sadness briefly passes across Brock's face only to be replaced with a smile and series of questions for the caretaker: "What happened to this one?"

In response to this question, we are told that a group of school children found the hyena inside their classroom one morning near Rumuruti. Their teacher called KWS, who then got in touch with Mount Kenya Wildlife Conservancy, where we now are, to rescue the cub. It is hypothesized that the hyena's mother gave birth to a litter close to the classroom during the night, and that this particular cub stumbled into the classroom. When the mother left the area before dawn, she was unaware of the cub she accidentally left behind. The story seems a bit unlikely for several reasons, but that is unimportant at this moment.

Photo 4.3. Author (Brock) holding a three-week-old hyena cub at Mount Kenya Wildlife Conservancy

"So, will it be released one day?" The response to this question gives us a temporary sigh of relief: "Yes, when it is maybe three years old. Some animals, like the buffalo and hyena, cannot be domesticated. At some point they turn wild, so we cannot keep them."

Our temporary relief comes from our need to believe that this hyena has a future outside the sanctuary. As lovely a setting as it is, nestled in the foothills and lush forests of Mount Kenya, we cannot help but question whether this space is a suitable long-term home for many of the animals held in sanctuary here. This doubt is fuelled by different scenes and situations encountered on our guided walk around the premises: a solitary baboon, rescued from the pet trade, that constantly paces back and forth, up and down, in its enclosure; a small group of Sykes monkeys separated by a chain-link fence from wild counterparts who routinely come to visit and interact with them; and a pair of cheetahs who "don't know" how fast they can run, according to the caretaker, simply because they never have run.

At the end of our tour, before exiting the compound, we stop and look back. The three-week-old hyena has been passed to another visitor, who has crouched down to let a large group of excited children see and stroke the animal. Further in the background, small children in large sun hats are taking turns climbing on top of Speedy Gonzales, a 150-year-old Galapagos tortoise, for photos. Our caretaker leads us to a sink inside an office where we can wash our hands before leaving. As we enter the building, he says, "We happen to be in the place where we accept donations for our work," pointing to a donation box. "Your support will help future animal orphans."

All the affective energies that had only just been coursing through our veins – of wanting to hold and be close to that baby hyena and be guaranteed of its happy future – begin to wash away, and our analytical minds kick back into gear. As we walk to the parking lot, we stop, place our own baby down on a recently cut lawn to play beneath the shade of flowering jacaranda, and immediately begin to debate what about our experience was and was not indicative of settler ecologies, and exactly what these types of spaces do for and to orphaned and injured animals. The drive down the mountain and back to Nanyuki is then spent questioning our own role in settler ecologies, and our approach to immersive research, which allows us to participate in situations such as these where we are simultaneously incredibly comfortable, due to our positionality and love for animals, and deeply uncomfortable, due to what we know about settler ecologies and how we have come to understand our place in them.

One recurring concern – of which there are many – we have had while writing this book is that we might be misinterpreted as lacking empathy for animal and plant populations that have been devastated by anthropocentrism and that are now bearing the brunt force of human-induced environmental change. This could not be further from the truth. Rather, we are in this line of work because of our deep affinity for nature. It is this affinity, along with our understanding of the colonial past and present, that leads us to question whether the approaches to saving nature that have become so dominant in Laikipia are effective and just. To be blunt: the conservation community lacks evidence that taking vulnerable, orphaned, or injured animals, like this baby hyena and Sudan the rhino, from their natural habitat and requiring them to do affective labour in exchange for saving their lives is the most effective and just model of species preservation. This particular model of conservation has gained traction for other reasons.

Fears about the biodiversity crisis and species loss are creating new geographies of wildlife conservation, including the emergence and expansion of spaces designed to prevent extinction, such as wildlife sanctuaries and rehabilitation centres. In this chapter, we traced the evolution of efforts to rescue and rehabilitate wildlife in Kenya, as they have evolved from a settler hobby and cultural norm to a full-fledged industry with ample national and international support and financing. We now finish by discussing how the rise of sanctuaries works to sustain structures of settler colonialism, in part by creating renewed demand for settler ecologists' expertise, land, and labour, while also nurturing, releasing, and retaining wild animals that demand their continued intervention.

In terms of expertise, because settler ecologists have dominated the rescuing industry for so many decades, they are often seen as de facto authorities on the rescuing and rehabilitation of wildlife – an image sustained by their propensity for publicizing stories of successful rescues in the news and on social media. As demonstrated throughout this chapter, when protected animals are found abandoned or injured, it is often settler-initiated organizations that are called in to do the work of rescuing. These organizations may work in close partnership with national wildlife authorities, but often it is the organizations' airplanes and vehicles that arrive on scene with their own staff and equipment to transport animals back to their own facilities for care and rehabilitation. Ultimately, this performance as experts in the rescuing of wildlife enables settler ecologists to remain highly relevant in a conservation landscape that, on paper, continues to shift further away from external and foreign control over the wildlife sector.

Alongside their expertise, settler ecologists have the land and resources required to establish and operate sanctuaries. Vulnerable and valuable species, like elephants and rhinos, cannot be placed and raised just anywhere after they are rescued. Large and secure landholdings are needed to keep these animals safe while connections to the international conservation community, including investors, donors, and researchers, are also paramount to keeping wildlife sanctuaries running and animals healthy. This is exemplified by the new rhino sanctuary on Loisaba, which has received finance and operational support from TNC, who recently purchased the conservancy. With access to land, capital, and connections that most others lack, settler ecologists are able to provide rare safe havens for species on the brink of extinction, which further secures their presence and influence in the landscape. Although there have been recent attempts to establish sanctuaries on community conservancies and inside national parks, even these tend to be operated in close partnership with settler ecologists.

Interestingly, departing from the maternal roots of the conservation-by-sanctuary movement, there has been a shift in the gendered nature of rescuing in recent years, which once positioned women as the primary carers of orphaned and injured animals. Today, the rescuing of wildlife is increasingly tangled up with the militarization of conservation, which involves the use of military-style approaches and technologies in the protection of nature (Lunstrum 2018; Verweijen and Marijnen 2018; Duffy et al. 2019). The newest generation of settler rescuers has often received training in the use of military-grade technologies and works closely with paramilitary organizations. They describe themselves as "poacher hunters" on social media, explaining how they pre-empt the need for rescue by catching and detaining trespassers and illegal hunters before they have a chance to act. Their images and words cast themselves as nature's heroes while rendering racialized African subjects as dangerous to nature (Lunstrum and Ybarra 2018), less-than-human (Mollett 2017), and part of a surplus population (Thakholi 2021) that can be sacrificed to save nature. By loudly narrating their role in rescuing species at risk of extinction – stories that are amplified by the international conservation community and documentaries, such as *Gardeners of Eden* – these settler ecologists have been able to carve out and secure a future for themselves in Kenya's conservation landscape.

A final way that the conservation-by-sanctuary movement contributes to sustaining settler ecologies is by nurturing quasi-wild animals that are at risk of demise without continued intervention. As discussed earlier in the chapter, wildlife sanctuaries generate revenue and moral support for their activities by facilitating intimate encounters between people and wild species. This includes spectacular experiences for tourists to interact up close with iconic species that would normally be seen only at a distance. These intimate encounters also extend to rescued animals and their caretakers. Wild animals raised, rehabilitated, and kept in sanctuaries exist in highly artificial settings where they are heavily reliant on and have regular interactions with caretakers – for example, being fed bottles and playing with the same caretakers for several years before going wild. These interactions are incredibly marketable, with rescuing organizations regularly telling stories online and to fee-paying guests about the special bond between particular orphaned animals and their surrogate human caretakers.

These quasi-wild animals are also incredibly valuable from a research perspective, as they offer guaranteed access to endangered species for research. Recognizing the research value of such facilities, epistemic institutions from all around have even contributed to financing rescuing facilities or acquiring land for research institutions, with Princeton University acquiring a former ranch in Laikipia to set up Mpala Research Centre and various zoological societies providing financing to rescuing facilities on Ol Pejeta. These financial arrangements, along with the knowledge produced through such research initiatives, highlight the incredibly important role that global epistemic communities play in sustaining settler ecologies.

As the conservation-by-sanctuary movement continues to grow, it is also important to consider the behavioural characteristics of quasi-wild animals and how these differ from their wild ancestors and counterparts. In cases where animals brought into sanctuaries are never released to the wild, their rescue places them in a state of permanent dependence on their rescuers. By establishing this dependence between wild animals and caretakers, rescuing facilities help secure their own futures. In cases where rescued animals being retained later reproduce, rescuing facilities are also able to use the growing population to claim more land, attract more financing, and exert more influence over conservation landscapes. Even in cases where animals can be released back into the wild, interactions between these animals and their caretakers can continue for many years. For example, it is common for orphaned elephants that have been released to return to their former stockades to visit their surrogate human guardians and friends and easily access food, water, and safety. Social media posts of those working for Sheldrick Wildlife Trust suggest that reintegrated elephants regularly spend significant periods of time at reintegration units and nearby eco-lodges in Tsavo and Chyulu Hills. This group of settler ecologists frequently posts photos of now wild elephant orphans in their pools, near their offices, around their vehicles, and being handled by staff members and visitors. Such interactions become even more common when these animals face hardship, such as conflict with other animals or during times of drought. This dynamic helps sustain settler ecologies by producing quasi-wild animals that demand ongoing intervention and contribute to the "spectacle" (Igoe 2017) of rescuing long after they are released.

5 Scaling

For the better part of a century, protected areas have been the predominant spatial unit of conservation for most states around the world – representing one of the first lines of defence in global efforts to protect biodiversity. Kenya describes its protected areas as "the jewels in [its] conservation crown" and sees these spaces as "represent[ing] the core elements of Kenya's conservation strategy" (GoK 2017, 2). Over the years, Kenya has developed an elaborate system of protected areas, including parks, reserves, conservancies, and sanctuaries in marine and terrestrial ecosystems. National parks and reserves are primarily managed by KWS in collaboration with national and county government authorities, whereas conservancies are primarily managed by private entities, communities, and organizations in collaboration with KWS. Although there are some KWS-managed sanctuaries, most are managed by other organizations. Over the past few decades, protected areas in the country have experienced notable growth, primarily through the expansion of private and community conservancies. Now numbering 160 individual landholdings, conservancies have gone from being a relatively uncommon land use to a form of protected area that covers 6.36 million ha or 11 per cent of Kenya's total land mass (KWCA, n.d.).

Despite the expansion of land for biodiversity conservation in Kenya, protected areas alone have been unable to halt biodiversity decline, habitat degradation, and species loss. This worrying trend is not unique to Kenya. While the world has over 260,000 protected areas, biodiversity continues to decline globally (Protected Planet, n.d.). One issue of mounting concern is that the protected-area model itself is ill-equipped to sustain populations of wildlife that require some degree of mobility on a routine and seasonal basis. In Kenya, species with notable ranges include elephants, Grévy's zebras, lions, and wild dogs. These animals are known to migrate regularly over fairly large distances in search of food, water, and breeding grounds. However, almost all wild animals adopt foraging behaviours and other consistent patterns of movement that could lead them outside protected areas on a routine basis. As a result, large proportions of wildlife end up spending significant periods of time each day, month, or year outside protected areas (Sindiga 1995; Wargute 2007; Western, Russell, and Cuthill 2009). When moving outside land areas strictly reserved for conservation, wild species are more likely to encounter humans and domestic animals that could threaten their safety and survival and vice versa. However, even species that might not migrate outside protected areas on a regular basis may experience hardship and harm within the boundaries of conservation areas. Because protected areas tend to be isolated from wider ecosystems and habitats, disease and parasite transmission, inbreeding, over-predation, and resource competition represent serious threats and hazards. All of these factors can lower survivability rates in protected areas.

Given how many protected areas exist around the world – each of which, to some extent, exists in a perpetual state of struggle to manage its own ecological challenges and constraints – calls are growing louder for a more connected approach to conservation that would allow wildlife to move between islands of protected areas. Wildlife corridors and dispersal areas are central to these calls for increased connectivity, and they are sometimes coordinated through the establishment of easements, community-conserved or wildlife-management areas, and engineered landscape features such as over- and underpasses. In this final chapter, we discuss the growing emphasis on ecological connectivity conservation (ECC) (Hilty et al. 2020) in Laikipia and adjacent areas. ECC seeks to secure habitat connectivity for wildlife by bridging different protected areas so that wildlife is able to meet its migratory and sustenance needs and increase its genetic diversity and resiliency (GoK 2017). In Laikipia, the uptake of ECC has catalyzed new efforts to reorganize land use and tenure systems on a large scale to create corridors and dispersal areas that allow wildlife to safely move in and out of protected areas.

In Laikipia, ECC is largely being pursued through the acquisition of land used for other purposes. By identifying and securing strategic parcels and plots of land and transforming them into safe and habitable areas, ecological networks are being stitched together to form bridges between protected areas in what otherwise resembles a mosaic of land uses. This approach aligns with the increasingly accepted wisdom that protecting biodiversity and saving species from extinction requires urgent action to connect habitat severed by human activities and forge safe passageways for wild species through human-dominated landscapes (Fitzsimons and Wescott 2005; Fitzsimons, Pulsford, and Wescott 2013; UNEP-WCMC 2018; Adams 2020). Such an undertaking necessitates a massive scaling up of conservation's territorial ambitions – away from grids of isolated protected areas towards planes of multiple, overlapping landscapes. We refer to this phase of ecological transformation as "scaling."

In this chapter, we draw attention to the expanding and increasingly diverse pool of settler ecologists involved in bringing about the transition towards ECC in Laikipia and beyond. Since Kenya's independence, KWS has played a key role in rewilding, repeopling, and rescuing, as have other relevant government organizations. However, to realize the scaled-up territorial ambitions required by landscape approaches to conservation, other governmental bodies have had to step to the fore. Recent years have seen the state take a keen interest in identifying, mapping out, and establishing wildlife corridors and dispersal areas across the country. This is exemplified by the government's new national plan, Securing Wildlife Migratory Routes and Corridors, which makes wildlife corridors and dispersal areas a pillar of Kenya's national development strategy, Vision 2030. Perhaps unsurprisingly, many nonstate actors have been supportive of this vision and instrumental in shaping the mechanisms, principles, and strategies behind it. Mount Kenya Trust, Space for Giants, TNC, and the UN Environment Programme represent just a small handful of the organizations involved in providing the funding, scientific expertise, and technical support needed to realize Kenya's vision for ECC. AWF, Save the Elephants, and Space for Giants, as well as other organizations engaged in conservation research, have also generated datasets that are being used to inform the land acquisitions needed for ECC.

Although the shift towards ECC is being driven by global trends and national and international conservation organizations, white settlers, in the traditional definition of the term, have also been central to this shift. In some respects, white settlers hold the key to realizing the scaled-up territorial ambitions of settler ecologists, broadly defined. In Kenya's national plans for ECC, settler landholdings are positioned as essential nodes within the larger networked conservation landscape that is under construction. Much of the land that will be involved in enabling wildlife to move with greater ease and security between private ranches and conservancies will be gained by enrolling settler properties in scaling initiatives. Furthermore, a few remaining wildlife-*un*friendly ranches in Laikipia could soon be secured as conservancies – providing the vital land bridges needed for wildlife to move with greater freedom through an expanded network of private ranches and conservancies.

The integral role existing settler landholdings have been assigned in Kenya's conservation connectivity framework means this latest phase of ecological transformation could further fix settlers in space and reinforce settler landholdings as strongholds of biodiversity conservation.

This chapter proceeds with a discussion of how protected areas went from being seen as the solution to the decline of biodiversity and loss of species globally to a problem in need of its own solution. The next part of the chapter consists of three sections that describe, in detail, three different approaches to ECC being pursued in Laikipia: the growth of community conservancies; the establishment of wildlife corridors; and a large-scale initiative called Ukanda wa Vifaru that seeks to create an ecological network for rhinos across much of Laikipia. Because landscape approaches to conservation are all about connectivity and mobility, this discussion invariably takes us out of Laikipia and beyond the fences of private conservancies to other spaces where species, such as elephants, migrate to and from Laikipia. In the final section before the book's conclusion, we also reflect on the increasingly diverse and expanding scope of settler ecologists being enrolled in efforts to realize the scaled-up territorial ambitions of ECC.

The Protected-Areas Problem

As protected areas became the primary spatial unit of biodiversity conservation over the course of the twentieth century, the ecological problems associated with enclosed national parks and reserves also became increasingly apparent. Many fortress-style protected areas in East Africa were proving to be far too small to sustain viable populations of wildlife. This is especially true of species whose behaviours and needs caused them to disperse into adjacent countrysides on a routine and seasonal basis, where they inevitably came into contact – and oftentimes conflict – with humans, domestic animals, and crops (Fryxell and Sinclair 1988; Western and Gichohi 1993; Kahurananga and Silkiluwasha 1997; Thirgood et al. 2004). As wildlife populations grew inside the confines of protected areas, some ecosystems also experienced degradation. For example, elephants can be incredibly destructive of their environments, especially when large numbers assemble in relatively small spaces, disturbing vegetation and water sources relied upon by other species. Rather than functioning as healthy, intact ecosystems, protected areas were turning into biogeographical islands where resource competition was growing, diseases were spreading, and inbreeding among certain species was rising (McClanahan and Young 1996). All of this lowered the survivability of wild species, threatening biodiversity, ecosystem health, and, in the case of zoonotic diseases such as anthrax (Stears et al. 2021), the health of domestic animals and people.

One solution to this problem pursued by governments and conservation organizations was the extension of existing protected areas and establishment of new protected areas to create greater space for growing populations of wildlife. Protected areas cover 27 per cent of land in East Africa, and new protected areas are continually being established, expanded, or upgraded across the region (Riggio et al. 2019). For example, Tanzania has created three new national parks, upgraded two game reserves to national parks, and expanded five existing national parks in the past two decades. Because the wave of new and expanded protected areas that swept across Tanzania at the turn of the twenty-first century involved the eviction and displacement of human populations, it was critiqued as a form of neocolonial green grabbing (Benjaminsen and Bryceson 2012). These patterns and trends can be seen on a much wider scale as well, across other parts of eastern and sub-Saharan Africa.

In Kenya, the growth of protected areas in recent years has primarily taken the form of private and community conservancies. It was hoped that the deficiencies of the protected-areas model might be addressed by accommodating novel spatial units of biodiversity conservation that would enable different actors, including individuals, private entities, and community trusts, to establish protected areas on their landholdings. Yet these new and usually geographically smaller protected areas are not immune to the very same problems as national parks and reserves. Furthermore, extending existing protected areas and establishing new conservation areas is not always seen as a desirable undertaking by governments and conservation organizations. Many protected areas are surrounded by human-dominated landscapes, including commercial and small-scale farms, grazing areas for livestock, and settlements. As can be seen in the case of Tanzania, expanding protected areas requires relocating humans and their livelihood activities, which is usually expensive and unpopular among communities with already-troubled relationships with protected areas and wildlife.

Another strategy for overcoming the ecological limitations and harms of protected areas is translocation, which involves moving animals from one protected area to another, as discussed at length in chapter 4. However, this too requires the existence of other protected areas within a reasonable distance capable of hosting more, as well as more diverse, species. Translocation is also a costly, labour-intensive process that can be dangerous for the animals being moved by aircraft or truck. For example, during a single translocation exercise in 2018, Kenya lost more black rhinos than it did to poaching in each of the previous two years (Save the Rhino 2018). For these and other reasons, translocation is rarely carried out on a routine, large-scale basis.

Thus, within the better part of a century, protected areas went from being an international solution for biodiversity decline and species loss to a problem requiring its own solution. The most recent ideas about how best to overcome the limitations of protected areas are reflected in landscape approaches to conservation. For several decades, scientists have been well aware of the challenges facing small, disconnected, and isolated wildlife habitats and have proposed various measures to create better-connected landscapes (Andrewartha and Birch 1954; MacArthur and Wilson 1967; Levins 1969). Although some of these concerns became evident within a decade or two of the protected areas movement in East Africa, they were taken seriously only at the turn of the twenty-first century. During this time, conservationists slowly began to acknowledge the need to shift away from isolated protected areas "towards areas with ecological value regardless of their territorial legal status, such as ecosystems, hotspots, ecoregions, bioregions, biospheres and landscapes" (Bluwstein 2021b, 5). Although protected areas still had a prominent role to play in larger conservation landscapes, they were no longer regarded as the most important spatial unit of biodiversity conservation.

Today, some of the world's most prominent international organizations are actively promoting and financing landscape approaches to conservation, including the IUCN and WWF (Ros-Tonen, Reed, and Sunderland 2018). In 2020, the IUCN was tasked with devising a coherent global approach for advancing landscape approaches to conservation, resulting in a set of global guidelines for establishing and widening ECC. More recently, ecological connectivity – defined as "the unimpeded movement of species and the flow of natural processes that sustain life on Earth" (UNEP 2020) – was positioned as a cornerstone of the Post-2020 Global Biodiversity Framework. Within this new global framework for managing nature, the connectedness of conservation landscapes is meant to serve as a key indicator for assessing how well countries are preserving and protecting nature in the midst of the biodiversity crises now perceived as global in scope (Díaz et al. 2019).

In response to shifting global conservation agendas, Kenya has recently demonstrated its commitment to landscape approaches to conservation by developing new national frameworks and legislation. In 2019, the government released its National Wildlife Strategy 2030, which calls for connected, resilient, and functional ecosystems. The government also published a national conservation connectivity framework, titled "Wildlife Migratory Corridors and Dispersal Areas:

Kenya Rangelands and Coastal Terrestrial Ecosystems." Around the same time, the chairperson of the Working Group on Wildlife Corridors and Dispersal Areas, Philip Muruthi, explained that the Government of Kenya no longer saw protected areas as sufficient in providing for the essential needs of wildlife. Rather, securing larger connected land-scapes for conservation was to be prioritized in Kenya's conservation action moving forward.

The Connective Conservation Solution

The landscape approaches to conservation being pursued in Kenya and advanced through the conservation connectivity framework place emphasis on identifying, coordinating, and strengthening ecological networks across great distances, spanning numerous land uses and, in some cases, multiple counties. Ecological networks are interconnected systems of core habitats, such as a cluster of conservancies or grouping of conservancies, national parks, and forest reserves, that can sustain much larger and more diverse populations of wildlife than isolated protected areas. One of the key ways that ECC departs from protected areas is that it allows wildlife to roam freely to access habitat, food, water, and a potentially more diverse gene pool for reproduction. Eco-logical networks accommodate the routine movements of wildlife on a daily or seasonal basis, while also offering individuals and groups the opportunity to move in response to local perturbations associated with environmental shocks or other hazards, such as conflict.

In the remainder of this discussion, we draw attention to three recent scalar initiatives that support Kenya's conservation connectivity frame-work and mirror the latest emphasis on landscapes in the Global Biodi-versity Framework (GBF). These initiatives include the creation of the following: community conservancies that connect the northern range-lands to core habitat areas in Laikipia, Mount Kenya, and the Aberdares; wildlife corridors strategically placed in highland areas to allow spe-cies, such as elephants, to navigate human-dominated landscapes; and a very recent connectivity initiative in Laikipia known as Ukanda wa Vifaru, or the Rhino Belt.

Community Conservancies

Under the umbrella of the NRT, community conservancies in Kenya's northern rangelands have come to represent the largest ecological net-work in the country. In chapter 3, we introduced the NRT and detailed the role the organization has played in incorporating pastoralism into settler ecologies and expanding settler ecologies into communal range-lands. In this section, we highlight other contributions the NRT has made to altering and reconfiguring ecologies in the rangelands. Specifically, we demonstrate how pastoral rangelands have become central to Kenya's vision and strategy for scaling up space for wildlife.

Since its formation in 2004, the NRT has arguably become one of the most influential actors in northern Kenya – so much so that the organization is likened to a "new" or "alternative government" by those who occupy space in the rangelands where it operates. As we discuss in chapter 3, "Repeopling," after Il Ng'wesi became the region's first community conservancy in 1995, several other communities entered similar arrangements with Craig, allowing their land to be used in ways that support wildlife conservation and, in some cases, leasing parts of their land to investors for tourism facilities in exchange for lease fees and jobs. Eventually, the NRT was formed to oversee all community conservancies in northern Kenya. In the years that followed, the organization experienced remarkable growth. By 2021, the NRT had established a total of forty-three community conservancies across eleven different counties in northern and coastal Kenya. These conservancies span 63,000 km² – covering over 10 per cent of Kenya's total land area – which is roughly the same size as Sri Lanka (NRT, n.d.; see photo 5.1).

The stated mission of the NRT is "to develop resilient community conservancies that transform lives, secure peace, and conserve natural resources" (NRT, n.d., para. 1). Initiated to help community conservancies better protect wildlife on their land, the organization adopted a broad mandate from its inception with the idea that improving the livelihoods and welfare of communities and building peace between conflicting groups would have positive impacts on rangelands and wildlife. For this reason, NRT initiatives cover a range of sectors and program areas: assisting with conservancy management, training, and fundraising; monitoring and evaluating conservancy performance; operating teams of rangers to monitor and protect wildlife and to combat other crimes, such as livestock theft and road banditry; promoting partnerships with the tourism sector; and running businesses to support livelihoods, such as a livestock-to-market and traditional handicraft enterprises. The NRT also has a Peace Team, which is responsible for managing, mitigating, and resolving conflicts within and outside community conservancies. Finally, the NRT provides guidance to conservancies on the management of rangelands for livestock and wildlife.

NRT conservancies are transforming ecosystems and altering ecological relations across the rangelands, not to mention in some parts of the coast and highlands as well. The details of these changes are

Photo 5.1. View of Il Ng'wesi and Lekurruki Community Conservancies after rains in 2023

documented in annual reports, as well as in a recently published study of the NRT's impacts on fauna and flora in conservancies between 2005 and 2019. This recent study highlights stabilizing and steadily increasing populations of several vulnerable and endangered species, including elephants, Beisa oryxes, and Grévy's zebras. Some species have experienced more dramatic increases in relatively short periods of time, such as reticulated giraffes, whose population increased from an estimated 1,509 individuals in 2012 to 4,019 in 2017 (NRT 2020a). According to the organization's "Status of Wildlife Report," NRT conservancies demonstrate significantly higher numbers of wildlife than communal lands without conservancies. However, in some cases, wildlife numbers are also now growing in unprotected areas adjacent to conservancies, such as the eland population in areas adjacent to Boni Forest in Lamu and Garissa Counties (NRT 2020a).

The NRT primarily attributes growing numbers of wildlife in and around its conservancies to three areas of intervention. First, the NRT argues that its efforts to improve habitat and grass cover have worked to the benefit of wildlife populations. The organization operates several different initiatives aimed at rehabilitating habitat, all of which are informed by the US Department of Agriculture's "Monitoring Rangeland Health Guide" and many of which are also supported by American funding and experts, such as the Department of the Interior and Geological Survey. Although livestock are permitted within NRT conservancies, conservancy-wide grazing plans control how many animals are allowed to graze in which parts of the conservancy and by whom. Additionally, the NRT clears invasive plant species and reseeds grass in its conservancies, with the goal of increasing perennial grass cover. For example, the organization has paid conservancy members to clear several thousand hectares of red-bark acacia – an invasive shrub that suppresses grass growth – and reseed them with perennial grasses. Several other restoration initiatives are run by the NRT as well, including healing eroded gullies by refilling them and using terracing and reseeding to prevent future erosion. Combined, these initiatives are meant to create landscapes that attract and sustain healthy populations of wildlife.

Second, the NRT argues that establishing new conservancies and improving security on these conservancies encourages wildlife to naturally expand their ranges into areas that were previously insecure and unsafe. The organization employs rangers to patrol conservancies and combat illegal activities, such as banditry and poaching for bushmeat and trophies. Many of these rangers are registered National Police Reservists, meaning they are permitted to carry and use weapons. To further improve security, the NRT has launched several specialized and highly trained security teams – each composed of ten NRT rangers and four KWS personnel. In addition to responding to incidents of road banditry and livestock theft, these teams are equipped to protect and monitor elephants and counter bushmeat hunting. A Joint Operations and Communications Centre (JOCC) is hosted at Lewa – serving as a regional hub for security and peace operations across all NRT conservancies. According to the NRT, low rates of illegal hunting, for example of giraffes, oryxes, and other animals consumed for meat, and very low levels of ivory poaching in conservancies illustrates the organization's success in protecting wildlife in communal rangelands shared by people and livestock.

Finally, the NRT believes wildlife populations are growing in its conservancies because it has created intact landscapes, providing connectivity and ranges that many species need to thrive, including cheetahs, leopards, lions, spotted and striped hyenas, and wild dogs (NRT 2020b, 6). The NRT uses data collected from aerial and ground surveys, as well as data that comes from GPS collars fitted to individual animals, to evidence how conservancies have transformed fragmented ecosystems into connected conservation landscapes. For example, between 2014 and 2019, the NRT partnered with conservation organization Save the Elephants to collar forty elephants "with a specific view to study the impact of conservancies in expanding safe range for elephants and identifying critical corridors in the landscape that must be conserved" (NRT 2020a, 18). Data generated from the collars suggested the ranges of elephants were expanding as new conservancies were established adjacent to already existing conservancies. Similarly, the NRT partnered with conservation organization Ewaso Lions to collar and collect data on lion movements between Samburu National Reserve, Buffalo Springs National Reserve, and several NRT conservancies, including Il Ng'wesi, Kalama, Lekurruki, and West Gate. The organization has used this data to demonstrate that community conservancies maintain the landscape connectivity needed to sustain large carnivores.

In addition to evidencing its role in creating connected, secure ecological networks, data collected by the NRT is being used to support the gazettement of new conservancies (NRT 2020a, 2020b). For example, the NRT has reported collecting data that demonstrates the recent and historical movement of elephants and black rhinos between Marsabit National Park, Meru National Park, and Lorian Swamp. This data was used to request US$10 million in funding from Rainforest Trust and Agence Française de Développement to create an ecological network between these protected areas by establishing four new community conservancies encompassing 3.2 million acres of land in total.

The massive exercise underway to scale up community conservancies into a vast ecological network spanning across the rangelands has been aided by the fact that the land conservancies are being established on has tended to be unregistered community land. Critics suggest this creates an opening for the NRT to "influence the apportionment of community lands for conservation purposes" to its own benefit (Noor 2019). However, the NRT is quick to argue it does not technically own conservancy land and, therefore, cannot be accused of dispossessing communities.[1] According to the NRT, it supports customary and communal land tenure security by helping "communities to register land for themselves" (Lalampaa 2021, para. 8). Regardless of who owns community conservancies in a technical sense, they are still spaces that are being reorganized and rezoned so that they benefit Kenya's landscape connectivity agenda. Diverse actors, ranging from Conservation International to KWS, the European Union to the San Diego Zoo, IUCN to USAID, and countless others, are working together to finance, create, and link new and existing NRT conservancies so that they provide connectivity for wildlife, thereby providing a solution to the protected-area problem.

Wildlife Corridors

In landscapes where protected areas are isolated because they are surrounded by land uses that cannot be as easily transformed for conservation, such as agriculture, infrastructure, and urban areas, wildlife corridors have become a popular strategy for removing barriers to wildlife movements and migrations. Corridors are strips of land that link healthy habitats, such as two existing protected areas, so that wildlife can safely move between protected areas to reach habitats that satisfy their needs. As Mara Goldman explains, "corridors offer a structural solution to the complex problem of maintaining functional ecological connectivity in a fragmented landscape by creating bridges that allow wildlife to move from one isolated habitat to another" (2009, 335).

---

1 In October 2021, the NRT published a series of Facebook and other social media posts aiming to debunk "myths" about the organization. Framed as a set of "True or False" discussions (or #FactsaboutNRT), the posts argued against growing criticism of the organization's complicity in land grabs, evictions, exclusionary policies, and other practices that work against pastoralists and pastoralism. This publicity stunt can be accessed via the NRT's Facebook page: https://www.facebook.com/NorthernRangelandsTrust.

As the Government of Kenya explains in its national policy on wildlife migratory corridors and dispersal areas, corridors are now seen as a pillar of the country's strategy for securing habitat connectivity, as they are understood to be the best approach for ensuring wildlife have "access to the larger habitat, while reducing inbreeding and improving genetic viability" (GoK 2017, 18). Corridors are also said to improve the safety and security of wildlife populations by "providing avenues for predation avoidance, while ensuring that essential ecological processes can continue" (GoK 2017, 18–19). Settler ecologists in Laikipia had already been experimenting with the use of wildlife corridors to create better connected landscapes well before the Government of Kenya released its national conservation connectivity framework. These early efforts remained somewhat ad hoc and small scale, and they predominantly had one species in mind: elephants.

Many of Kenya's early corridor projects were geared towards helping elephants move from one protected area to another in increasingly fragmented landscapes. The early focus on elephants was partly due to the recognition that elephants embark upon far more extensive movements and migrations than most other land mammals. By the 1990s, it had become the accepted wisdom that elephants made use of multiple home ranges linked together by migratory corridors. It was widely believed that elephant corridors were ancestral – with elephants passing on mental maps of their routes from one generation to the next – and, as such, there was mounting concern that industrial expansion and urbanization were interfering with elephants' intergenerational knowledge of suitable migratory routes. In response, early pioneers of the corridor movement in Kenya, such as David and Daphne Sheldrick, began attempting to reopen and secure elephants corridors they identified as still existing within the living memories of specific herds.

In Laikipia, Mike Prettejohn, owner of Sangare Ranch, was among the first settler ecologists to leverage growing interest in elephant corridors into a concrete intervention. In the late 1990s, Prettejohn proposed the reopening of a corridor that would allow elephants to move across a densely populated and settled area between the national parks of Aberdares and Mount Kenya. Prettejohn's property sat directly between these two national parks. According to Prettejohn's reading of history, early colonial settlers in this area remembered large herds of elephants moving between the parks, regularly traversing what was now his property. For example, he had read that Captain Richard Meinertzhagen, a British officer serving with the King's African Rifles, witnessed a procession of 700 elephants travelling this route in the early 1900s (Graham 2001). From Prettejohn's perspective, reopening this safe passage for elephants was essential for restoring the natural movement of elephant populations in the region and ensuring their future health and survival (Graham 2001). Prettejohn proposed that smallholders who had settled in this ancient elephant corridor should be relocated and compensated to reopen the corridor. Naturally, Prettejohn did not advocate for his own eviction, framing his property as wildlife friendly and a safe haven for elephants travelling between Aberdares and Mount Kenya. Perhaps unsurprisingly, smallholders targeted for relocation saw the corridor project as a "cover-up for whites taking over African land" (van den Akker 2016, 142), and the project was all but abandoned by the early 2000s.

After several years of silence, conversations about the Mount Kenya-Aberdares Elephant Corridor have recently been reignited. Now an initiative of Nyeri County government, KWS, and conservation organization Rhino Ark, plans have been revitalized to establish a 25 km fenced elephant corridor connecting the two parks as initially planned. So far, a study has been undertaken to assess the corridor's feasibility, along with consultations and workshops to persuade existing landowners about the benefits of the corridor (ATC News 2018). Although the proposed route for the corridor is now far more blocked than it was before – by both land subdivision and the Nairobi-Nanyuki highway – plans to move forward with the corridor are now making headway.

Not long after the initial proposal for the Mount Kenya-Aberdares Elephant Corridor was abandoned, the Mount Kenya Elephant Corridor was proposed and implemented with far more immediate success. In 2006, two settler landholdings on the north-facing side of Mount Kenya, Kisima Farm and Marania Farm, joined forces with Lewa and two conservation organizations, the Bill Woodley Mount Kenya Trust (BWMKT) and the Ngare Ndare Forest Trust (NNFT), to establish a wildlife corridor connecting Mount Kenya's northernmost forests with the adjacent Ngare Ndare Forest Reserve and adjoining Lewa Wildlife Conservancy. The farmers involved in the project hoped the corridor would help to prevent crop destruction on their own properties by directing the flow of elephants in and out of the forests. This collective of settler ecologists saw the corridor as an opportunity to provide elephants with better protection against poaching and human-wildlife conflict while also contributing to the diversification of the elephants' gene pool by enabling elephants from the lowlands and highlands to mix (van den Akker 2016).

Unlike the initial elephant corridor proposed further south by Prettejohn, the Mount Kenya Elephant Corridor was designed to avoid the eviction and relocation – or any involvement – of smallholders. Instead, the corridor was planned to traverse through conservancy land and private property, beginning at Mount Kenya, running along the outer boundaries of Kisima and Marania farms, and ending at Ngare Ndare Forest Reserve. From Ngare Ndare Forest Reserve, elephants would have secure access to the Lewa-Borana Landscape, key habitat areas in Isiolo and Samburu contained within national reserves, and a vast network of NRT conservancies. Even though acquisition of land for the corridor was relatively straightforward, the project still experienced challenges, primarily around the US$1 million needed to construct and maintain the corridor, which required an underpass beneath the A2 highway between Nanyuki and Meru and a 14 km elephant-proof fence (Nyaligu and Weeks 2013). However, funding was ultimately secured from Richard Branson's Virgin Atlantic and the Embassy of the Kingdom of the Netherlands (van den Akker 2016).

Construction of the Mount Kenya Elephant Corridor began in 2008 and was finished in 2010, with elephants reportedly using the corridor within just one week of its opening. With the opening of the corridor, the population of elephants in Mount Kenya National Park – approximately 3,700 individuals at the time – was linked to a population of 500 elephants in the Lewa-Borana Landscape and over 2,000 more elephants in the lowlands of Isiolo and Samburu (Litoroh et al. 2012; Green 2016; Kimiti, Mwololo, et al. 2017). Within this expanding elephant landscape, the growth and health of herds in the past few years has been used as evidence that the Mount Kenya Elephant Corridor was successful in enabling elephants to thrive in what was otherwise a fragmented landscape. There is also growing evidence that creating greater connectivity for elephants in the region has supported the mobility and migratory needs of other species as well (Didier et al. 2011). Studies have found that several species have made the Mount Kenya Elephant Corridor a permanent part of their home range and foraging areas (Winmill 2014; Green 2016).

As we have discussed in relation to other phases and approaches to ecological transformation in Laikipia, the creation and expansion of ecological networks is not strictly an ecological intervention; it supports other economic, social, and political interests for settler ecologists. A particularly telling outcome of the Mount Kenya Elephant Corridor in this regard was the extension of Mount Kenya National Park World Heritage Site to include Ngare Ndare Forest Reserve and Lewa Wildlife Conservancy. The UNESCO World Heritage Committee first recognized Mount Kenya as a Natural World Heritage Site of Outstanding Universal Value in 1997. In principle, this designation is meant to bring a range of social and economic benefits to an area, including access to funds and increased legal protection and international recognition. Following the construction of the Mount Kenya Elephant Corridor, Lewa submitted an application requesting the Natural World Heritage Site be extended to include the new land areas now directly attached to Mount Kenya National Park via the Mount Kenya Elephant Corridor. UNESCO approved the application in 2013, deeming the ecological network of Mount Kenya, Ngare Ndare Forest, and Lewa as significant enough to extend the existing heritage site.

Critical onlookers suggest that, from its inception, the Mount Kenya Elephant Corridor was about far more than conservation connectivity. As van den Akker (2016) argues, through the corridor project, Lewa was ultimately granted UNESCO World Heritage status, further securing the presence of settlers, settler ecologists, and settler ecologies in the landscape. That Lewa is now seen by the world as a global natural treasure symbolizes just how powerful and significant settler ecologies are in the eyes of the international conservation community. Van den Akker (2016) reports that this status was pursued as further security against any potential future land reform program in Kenya, which is a sentiment we heard from other sources. Following Lewa's success in securing UNESCO World Heritage Status, other settler ranches and conservancies are now looking for similar recognition to help secure their property rights, with Borana at the forefront of this movement. Citing a former project officer for Laikipia Wildlife Forum, Gitau Mbaria writes, "There is a rush to create a super big protected area stretching from Lewa to Solio – all of it under the cover of World Heritage Convention" (2017, para. 19).

Ukanda wa Vifaru

Ukanda wa Vifaru (the Rhino Belt) is another pertinent example of the shift in emphasis away from establishing isolated protected areas to creating connective conservation landscapes. Ukanda wa Vifaru is a proposed rhino landscape encompassing 544,000 acres that aims to allow rhinos to move securely between four existing rhino sanctuaries, Ol Pejeta, Ol Jogi, Borana, and Lewa, and one new rhino sanctuary, Loisaba (see map 5.1). Should the full project be realized, it would create an ecological network for rhinos that extends from the south of Laikipia to the north of the county, near Oldonyiro, to Borana and Lewa in the northeast. Unlike the ecological network created by NRT conservancies on communal and trust land, Ukanda wa Vifaru would create a network primarily consisting of private ranches and conservancies – with the exception of Il Ng'wesi, Ngare Ndare Forest, Mutara Conservancy, which is jointly managed by Ol Pejeta and the Agricultural Development Corporation, and the proposed Laikipia National Park, which is currently trust land. Cutting across the old ranching heartland of Laikipia, most of these properties are large settler landholdings that already serve as an ecological network for some species. Although many species already move between these different properties, the movement of rhinos is restricted outside dedicated sanctuaries due to the risk of poaching. By connecting Ol Pejeta, Ol Jogi, Borana, and Lewa, Ukanda wa Vifaru is intended to create a much larger, secure migratory and dispersal area for rhinos that will also benefit other species.

Due to the mosaic of land uses where Ukanda wa Vifaru is being proposed – including ranches, mixed-use ranches, private conservancies, community conservancies, trust land, and a forest reserve – the initiative has been broken down into three phases. Phase 1 is now complete (Helping Rhinos, n.d.) and has involved the preparation of two properties adjacent to Ol Pejeta for rhinos. Phase 1 is considered the most complicated, as the two properties that need to be secured were not previously considered wildlife friendly.

To bring these two properties into the rhino landscape, both have undergone land acquisitions and management changes. To start, in 2008, the AWF and TNC joined forces to purchase one of these properties, Eland Downs – previously owned by former president of Kenya, Daniel Arap Moi. Both the AWF and TNC hoped to transform the 17,000 acre farm into a new protected area, Laikipia National Park. At the time of purchase, the land was being used by several different communities for farming and livestock keeping. During the land transfer, somewhere between 2,000 and 3,000 families were violently displaced, with the AWF and Kenya Police Reserves implicated in the evictions. Following public outcry, the AWF and TNC attempted to lessen their involvement in the messy situation by gifting the land to KWS with the intention of partnering with the government to manage the new national park after its establishment. Unsurprisingly, returning the land to the government failed to appease the evicted families and the case remained tied up in court for many years.

More recently, in 2018, Ol Pejeta secured the second of the two properties involved in phase 1 of the project, Mutara Ranch (now Mutara Conservancy). Mutara Ranch was also held by the government at the time of purchase and managed by the Agricultural Development Corporation. Possibly learning from past mistakes, local communities were not evicted from this 20,000 acre parcel of land upon acquisition. Instead, Ol Pejeta committed to working with the communities living on the

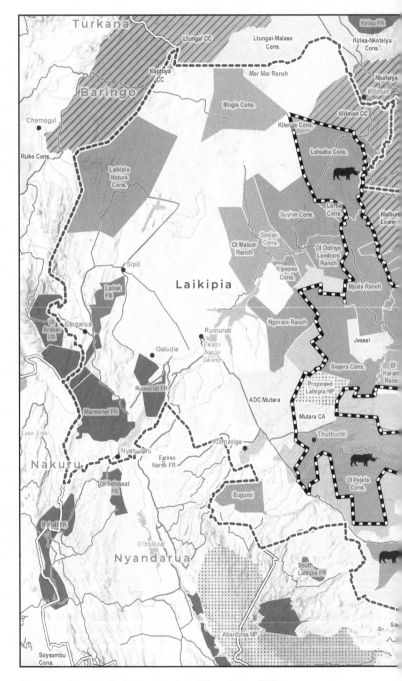

Map 5.1. Proposed rhino corridor, Ukanda wa Vifaru

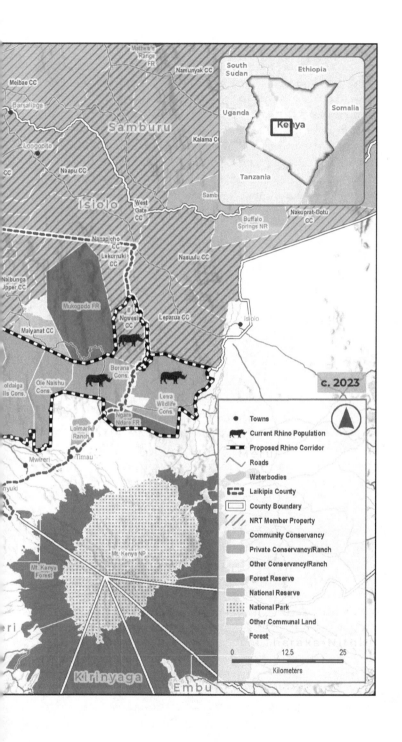

property to set aside certain parts of the ranch for conservation while also restoring grazing lands for cattle, establishing a fodder production system, and developing an LTM initiative. Ol Pejeta also rehabilitated four miles of pipeline across the ranch, providing improved water access for people, livestock, and wildlife. Since 2018, habitat health on Mutara has significantly improved and there has been a sharp increase in the number and distribution of wildlife across the ranch (WildLandscapes International, n.d.). For example, in 2019, a pride of lions permanently migrated from Ol Pejeta to Mutara and elephants now frequently move between the two conservancies. Given how quickly wildlife was attracted to Mutara after ecological conditions were transformed, proponents of the rhino corridor believe habitats on the ranch should be healthy and secure enough to be opened up to rhinos from Ol Pejeta in the near future.

Plans for phase 2 of Ukanda wa Vifaru have also recently begun to take shape. Phase 2 will involve preparing three existing wildlife-friendly properties within the ecological network to host rhinos: El Karama, a mixed-use ranch with tourism facilities and activities owned by a settler family; Segera, a ranch with wildlife tourism activities owned Jochen Zeitz, a German philanthropist who sits on the board of trustees for KWS and was formerly CEO of PUMA; and Mpala, a massive wildlife research centre owned by Princeton University. Together with Mutara Ranch and Laikipia National Park, these five properties will form a land bridge connecting the rhino sanctuaries of Ol Pejeta, Ol Jogi, and eventually Loisaba. As El Karama, Segera, and Mpala are already wildlife-friendly properties, preparing them to host rhinos is less about habitat restoration and more about security upgrades. In addition to more investment in fencing, rangers, and surveillance, these properties will need to raise significant funds to cover operating costs if they are to host rhinos. By some estimates, an average of US$500,000 is needed each year to support the security costs of hosting a viable rhino population. Interestingly, however, just a few years ago, a manager of one of these properties said to us in conversation, "No, we'll never have rhino. Too expensive." It may be that novel forms of conservation finance and investment have changed this manager's tune. For example, WildLandscapes International, a US-based organization dedicated to conserving large and connected landscapes through public-private partnerships, has agreed to help raise US$25 million to support El Karama, Segera, and Mpala in carrying out the necessary upgrades and preparations required to operate as rhino sanctuaries.

After phases 1 and 2 have been completed, phase 3 of Ukanda wa Vifaru will focus on preparing a further four properties for rhinos:

Lolldaiga and Ole Naishu, which are both settler-owned cattle ranches with tourism facilities and activities; Enasoit, a private conservancy owned by a Norwegian family that put down roots in Kenya three decades ago; and Ngare Ndare Forest Reserve, which is part of the Mount Kenya World Heritage Site and is managed by extension in partnership with Lewa Wildlife Conservancy. As with phase 2, innovative funding mechanisms are being pursued to help raise the capital needed for these remaining properties to function as secure rhino habitat. For example, it is hoped that Rhino Bonds will provide the investment needed for Ngare Ndare Forest Reserve. Rhino Bonds work by attracting private investment in preparing land for rhino conservation with the promise of financial returns if rhino numbers increase over a five-year period.

One issue that could impact phase 3 of the project is that some of the remaining landholdings have plans to sell in the coming years. It has been suggested that WildLandscapes could help secure the capital needed to purchase these properties so that they can be "managed as a continuous conservation corridor in perpetuity" (WildLandscapes International, n.d., 11). Should this happen, ownership of these properties may be transferred to existing conservancies, such as Lewa. Alternatively, Laikipia Forum has proposed the establishment of a new Conservation Land Trust that would "protect key parcels of the landscape, and help to ensure the contiguity and connectivity of rangelands for wildlife" (Laikipia Forum, n.d., para. 4). The trust would be "designed to offer landowners the choice to leave their land in conservation status for perpetuity" in the event of their death (Laikipia Forum, n.d., para. 4), reflecting the permanent and enduring intentions of settler ecologies in this landscape.

Ukanda wa Vifaru occupies a key position in Kenya's National Black Rhino Action Plan (2017–21), which is committed to identifying areas for black rhino population expansion to achieve the strategic vision of at least 2,000 black rhinos in Kenya (Khayale et al. 2020). Black rhino numbers are already growing in the highly secured sanctuaries of Laikipia, but many of these sanctuaries have either reached or are approaching the limit when it comes to the population sizes of rhinos within their properties. Because rhino sanctuaries safeguard rhinos from predation, shield them from competition for food or depleting water sources, minimize exposure to zoonotic disease, and provide medical attention when needed, populations of rhinos have grown far quicker in sanctuaries than they would in more natural conditions. When rhino populations become too large relative to the habitat available, it is not just competition for food that becomes a problem; rhinos are also at increased exposure to diseases and parasites. The government understands that achieving its strategic vision for rhinos requires looking beyond the expansion of existing sanctuaries or creation of new secure enclosures to the level of landscapes. Ukanda wa Vifaru is an example of the connective conservation models required for rhino populations to continue growing and expanding (Khayale et al. 2020).

Importantly, the ecological implications of an initiative such as Ukanda wa Vifaru extend far beyond bolstering rhino populations alone. As WildLandscapes explains, "Ukanda wa Vifaru will protect and restore nearly 544,000 acres and link an ancient wildlife corridor from the heart of Laikipia to the arid lands of the north, providing a total of 7.5 million acres of landscape connectivity for wide-ranging species" (n.d., 15). Unlike most other species, rhinos in Kenya are deterred from roaming freely. They are required to be contained within secure protected areas that offer 24/7 surveillance. Within these spaces, extensive security and monitoring systems are used to prevent poaching, such as innovative ring-fencing systems with motion-triggered camera traps that make it difficult for anyone unwanted to enter (see photo 5.2). Even if someone manages to make it through a fence, they will likely soon be apprehended by dedicated teams, composed of armed and highly trained rangers that patrol by foot, vehicle, and aircraft.

When security requirements are considered alongside issues such as population size and habitat availability and quality and financing, questions must be raised about whether there is any way this ecological network will not be highly exclusionary. For example, to what extent will people who are not conservationists or tourists be free to move within or across Ukanda wa Vifaru? While pastoralists are currently permitted to enter community conservancies such as Il Ng'wesi to graze and water their livestock, will this change if rhinos are to be afforded the ability to move across the 544,000 acres being put aside for Ukanda wa Vifaru? Will LTM initiatives in private ranches and conservancies currently without rhinos remain an option for herders and livestock when Ukanda wa Vifaru is realized?

As discussed in chapter 4, only a handful of properties in Laikipia have been able to meet the stringent and expensive requirements of protecting rhinos on their land. However, Ukanda wa Vifaru stands to change this. Even as some settler families within the proposed footprint of the ecological network are considering selling their land, plans are also being made and mechanisms put in place to transfer ownership to global conservation organizations – a different type of settler ecologist. The ultimate aim of this scaling exercise is not just to expand corridors,

Photo 5.2. One of the properties within the proposed rhino corridor

networks, and connective conservation landscapes, but also to secure settler ecologies on a grand scale in perpetuity (WildLandscapes International, n.d., 11).

Connective Conservation and Its Immobilities

To understand the implications of landscape approaches to conservation being advanced in Laikipia, it is important to look beyond the fences of private ranches and conservancies to spaces outside these protected areas that serve as dispersal areas and migratory corridors for wildlife. It is also necessary to consider the experiences and knowledge that people in these spaces hold of wildlife behaviours and migratory patterns on their land and how these behaviours and patterns are affected by interventions made by settler ecologists in pursuit of ecological connectivity. Before concluding this chapter, we do just this, briefly drawing attention to some of the ways pastoralists on the Il Ng'wesi side of the Lewa-Borana Landscape have observed and experienced changes in animal mobilities associated with recent conservation connectivity interventions.

Photo 5.3. The fence gap at Arijiju that allows wildlife to move throughout the landscape but prevents rhinos from leaving Borana Conservancy and Lewa Wildlife Conservancy

Late one afternoon in July 2019, we stand with two elders in the middle of Arijiju Gap, an opening in Borana's fence that is meant to allow all wildlife aside from rhinos to come and go from the conservancy (see photo 5.3). We approached the gap on foot from the Il Ng'wesi side, after spending the day with elders from the community of Lokusero learning about the movements of livestock and wildlife in the landscape, as well as strategies adopted by pastoralists in the area over time to enable coexistence. Leaning on short poles positioned in the gap that serve as obstacles to prevent rhinos from escaping, we listen to the elders explain that multiple gaps, including Arijiju, were placed in the Borana-Lewa fence without the consent of communities in Lokusero or Sanga, two settlements in Il Ng'wesi located near the gaps. In the process, other migratory routes known to the elders and other herders from their communities were blocked, forcing wildlife to pass near or through settlements and to spend more time around homesteads and livestock enclosures as they look for a way to get around or through conservancy fences.

The previous week, a different elder showed us evidence of migratory routes near Seiku Gap, now blocked by the Borana-Lewa fence. After observing the passageway that had been built into Borana's fence, we walked for a kilometre or so westward to a place where a gully had been dug out beneath the fence, creating a shallow opening. "Here, a male lion comes one or two times per month. It goes down into the conservancy [Il Ng'wesi] there searching for lioness," the elder explains, pointing away from Borana Conservancy in the direction of the Il Ng'wesi's eco-lodge. The lion's claw marks are visible in the hard, dry soil beneath the fence, and dried faeces nearby serve as further proof of the big cat's presence. "I have watched it pass just there," he says as he points northward across the road. "My house is there" – he nods towards a *boma* in the other direction. "At night, when the lion roars, you can feel it here," he says, putting a hand on his chest.

After demonstrating how the lion drops down into the gully and shuffles under the fence on its belly, the elder crosses back to the Il Ng'wesi side of the fence and begins to relay lessons he learnt as a child about respecting wildlife migratory routes and accommodating the movement of wildlife in the landscape. "We are brought up to know that we should not block wildlife migratory corridors when we take our livestock to graze and wildlife is always left to drink from their wells in the night-time and to graze on their pastures unencumbered near us." Although not using the language of ECC, as we do in this chapter, this elder was sharply critical of efforts to create intact landscapes relying on private land regimes, new infrastructure, and partial knowledge of how ecosystems and ecological relations within these landscapes function. When asked if residents of Lokusero and the surrounding era were consulted about how fences were constructed or where fence gaps were positioned, the elder replied, "Never. And I do not know anyone from here that was consulted. Look, they have put the corridor straight into our community. Truthfully, it should be nearer Ngare Ndare [River]," because that is where the animals historically crossed.

We spent a number of weeks with other elders across Il Ng'wesi who relayed similar lessons imparted upon them when they were young herders. All the elders also spoke at length about their knowledge of behaviours and migratory patterns associated with different species, as well as past and present strategies for maintaining flows of wildlife, livestock, and people across these shared lands. Like that of the elder at Seiku, one of the common messages was that by disrupting flows of wildlife, the fences and gaps constructed around conservancies were also undermining ecological connectivity and coexistence. Historically, coexistence on lands shared by humans, livestock, and wildlife allowed for the optimization of ecological benefits for a wide range of species. For example, herds of cattle and elephants shared migratory corridors directing them to shared reserves of food and water. Their regular movement along these corridors helped maintain clear passageways through bushes, forests, and rocky escarpments that were otherwise difficult to discern and navigate. Elders perceived dominant approaches to reorganizing and reordering landscapes as upsetting these ecological relations, making the behaviours of people and wildlife less predictable to each other and replacing relations described by elders as once "respectful" and mutually beneficial with those now perceived as "disrespectful" and "aggressive."

Because elephant densities in what is now Il Ng'wesi were far lower historically, herders could also climb up to strategic viewpoints overlooking the grass and shrublands below to coordinate their daily movements with any sign of elephants or other potentially hazardous animals, such as buffalos and lions. As a general rule, which did not always apply in the case of rogue elephants or during mating seasons, nonpredatory species also did their part to avoid close contact with humans and cattle. As the two quotations from different elders below suggest, this was before the time when modern, electrified fences were strung across the landscape and new conservation and tourism activities encouraged species such as elephants to seek out human presence.

> Fences have blocked [wildlife] movements and now they just sit along the fence ... Fences are blocking the movements and routes of elephants, so now they are staying in places they would never stay in before. Importantly, however, it's not fences alone that are blocking the movements and routes of elephants. It is also because the fences are more modern – they have been electrified.

> The problem is, there are too many fences ... These fences are blocking these wild animals from leaving the community so we stay with them in a very small area. If there are no fences, we would stay very well with [wildlife] because [they] come to the community where people live during the night and go far during the day. Fences have changed this behaviour.

These quotations reveal the contradictions inherent in scaling and the disregard for ecologies otherwise that do not rely on fences and fence gaps, security teams and surveillance systems, when pursuing and maintaining connected landscapes.

With all this in mind, the ecological implications of scaling require careful monitoring and ongoing scrutiny. As this chapter makes clear, the connective conservation solution being rolled out across Laikipia and in its surrounding environment relies heavily on the acquisition of new land and securitization – not just of private ranches and conservancies, but of entire landscapes encompassing varied types of landholdings. Fences are not being removed – they are being lengthened – and the amount of land within these expanded fence lines is being scaled up. As the territorial ambitions of ECC are realized through this approach, equally vast expanses of ecologies otherwise and potential ecologies otherwise are being subsumed and exposed to the risk of erasure.

Moreover, if the vision of landscape connectivity for wildlife across the Samburu-Laikipia Ecosystem is achieved as it is currently imagined, it could mark a return to a period during the early colonial era where the landscape was highly segregated. The vision of Ukanda wa Vifaru in its current form, as well as other connective conservation initiatives, risks ushering in an era of *ecological apartheid* where divisions between different groups – including different groups of humans, as well as nonhumans – are extremely stark (Büscher 2021; Koot, Büscher, and Thakholi 2022). As Bluwstein (2021a) writes, landscape-based conservation initiatives are deeply interventionist and often coercive: making these initiatives a reality results in more, rather than less, bordering of conservation areas and separation between people and nature (see also Noe 2019; Ramutsindela 2007).

Making Settler Ecologies: Scaling

Concerns about the worsening global biodiversity crisis are supported by grim statistics: one million plant and animal species are now threatened with extinction (Díaz et al. 2019); over 500 species of land animals are likely to be lost within twenty years (Ceballos, Ehrlich, and Raven 2020); and more than 1,400 species of trees are critically endangered and in urgent need of conservation (FAO and UNEP 2020). The decline of fauna and flora at such a rapid rate and global scale has led some scientists to claim the world is now experiencing another mass extinction event – a period of time when species start to vanish faster than they are replaced (Ceballos et al. 2015).

Although climate change sits at the forefront of explanations for what is causing such a significant decline in biodiversity and loss of species, there is another narrative that the international conservation community also relies upon to explain the origins of the biodiversity crisis. This narrative suggests the growth of human populations – which goes along with greater numbers of livestock, the conversion of land for human uses, and the expansion of settlements and urban areas – is also a primary driver of species loss (Weldemichel 2020). According to overpopulation theories, activities associated with the growth of human populations contribute to changes in land use and vegetation cover that result in habitat fragmentation and degradation, which contribute to declines in biodiversity.

Those who rely primarily on overpopulation narratives to make sense of biodiversity decline argue that world leaders need to urgently scale up conservation action. In particular, the idea that much more of the earth's surface should be set aside for biodiversity and protected from humans is gaining momentum. One proposal, known by the slogan "nature needs half," spearheaded by conservation scientists such as Wilson (2016), Noss et al. (2012), and Wuerthner, Crist, and Butler (2015), suggests that 50 per cent of the earth's land- and seascapes need to be shielded from human uses to stave off "a mass extinction unrivalled in the last 65 million years of life on Earth" (Cafaro et al. 2017, 400; see also Secretariat of the Convention on Biological Diversity 2014). This protected-areas solution, which is being promoted through the Half Earth Movement, proposes scaling up the world's network of protected areas by expanding existing reserves and improving connectivity to create a grid of protected spaces encompassing half the world's surface.

The Half Earth Movement has been critiqued for being unrealistic and unjust (Büscher et al. 2017), but a somewhat more moderate proposal has been accepted through the GBF. Known as "30 by 30," this plan proposes protecting 30 per cent of the earth's land, oceans, and freshwater by the year 2030. Unlike the Half Earth Movement, the 30 by 30 agenda adopts a broader definition of what counts as protected land, with the aim to conserve 30 per cent of the earth's land and water through formal protected areas, as well as through other effective area-based conservation measures (OECMs). Private conservation areas, wildlife corridors, community-managed forests, and sustainable fisheries will contribute to this 30 per cent target. With the GBF now ratified, the 30 by 30 target will shape the next decade of global biodiversity conservation policy.

Agendas such as Half Earth and 30 by 30 are taking the spatial visions, ambitions, and practices of conservation to levels never before imagined (Adams 2020), and are both shaping and being shaped by the trend of scaling that we describe in this chapter. Renewed global emphasis on expanding territory for conservation is providing impetus for landscape-level interventions to take shape in Laikipia. In recent years, the creation of ecological networks and connective landscapes have come to dominate conservation policy and discourse in the region. The establishment of wildlife corridors between protected areas is one aspect of this new connective conservation framework, as is the assembly of wildlife-friendly properties to expand dispersal areas and habitat ranges for all sorts of species. Importantly, as we show in this chapter, concerted and coordinated scaling efforts are not only spatial – they are also temporal. New financing arrangements and legal mechanisms are being devised to secure connective landscapes in perpetuity.

Problematically, narratives that serve as a rallying cry against the ills of overpopulation and for the advancement of landscape approaches to conservation risk stigmatizing some rural land users – human and nonhuman – and land uses (Cavanagh, Weldemichel, and Benjaminsen 2020, 1594). As Kenya's connective conservation framework states, "the diminishing wildlife ranges is a result of human pressure on land resources and is aggravated by land tenure changes, agricultural encroachment along rainfall gradients in the rangelands, and high density settlement and livestock numbers in key wildlife dispersal areas" (GoK 2017, 7). Such discourses are used to justify rearranging and reordering entire landscapes – creating hierarchies of animals, plants, humans, and ecological functions – while fixing agriculture, pastoralism, and other land-based livelihoods in space – squeezing populations considered surplus or threatening into interstitial and often marginalized areas (Bluwstein 2018; Cavanagh, Weldemichel, and Benjaminsen 2020).

Wildlife corridors drawn around human settlements – often without input from those holding intergenerational, tacit knowledge of wildlife behaviours, migratory patterns, and foraging needs – demonstrate how rural populations risk being fixed in space by landscape approaches to conservation. In addition to physically constraining the movements of pastoralists, herders, and livestock, these new approaches may inhibit them from contributing to the maintenance of shared corridors and other ecosystems that benefit from coexistence. As maps of new conservation landscapes are produced, and new corridors, dispersal areas, and networks are marked out on these maps, pastoralists are also alienated ideologically from the biodiverse landscapes they previously helped produce and maintain. This leads to a kind of ecological apartheid, with stark separation enforced between mobile populations of people, domestic animals, and wildlife (for more on green or ecological apartheid, see Koot, Büscher, and Thakholi 2022).

As scaling conservation landscapes risks fixing smallholders and pastoralists in space, it could also result in the securing of more space for settler ecologists. In such a tightly packed mosaic of land uses, it is significant that new connective conservation initiatives and financial and legal mechanisms appear to be freeing up land for settler ecologies. Scaling exercises are repositioning private ranches and conservancies as strongholds of biodiversity conservation and beacons of the future. During an interview with us in 2016, a former executive director of Laikipia Forum – which is behind the Conservation Land Trust idea – pronounced,

"The land tenure issue? That's been resolved in Laikipia!" However, the land tenure question in Laikipia is anything but resolved. A new scramble for land tenure is underway and, this time, the goal is to secure the future of the entire landscape in a more lasting way than in the past.

By no means are white Kenyans the only actors behind the scaling up of conservation in Laikipia, nor could their efforts alone make scaling possible. A range of other nonstate actors are involved in the process. Most pertinently, without finance from international conservation organizations and social impact investors, much of this work would be infeasible. Consultants and researchers are also involved. They collect data used to evidence how wildlife move about the landscape, as well as what and who act as barriers to mobility. Data from this class of experts is used to map, acquire, and consolidate land to extend the territorial claims of conservation (Bluwstein 2018) and to ensure connective conservation landscapes appear ecologically and economically viable (Cavanagh, Weldemichel, and Benjaminsen 2020). Finally, this work would all be in vain without the backing of state actors, like KWS. The expanding territorial ambitions of conservationists are animated and emboldened by complementary state agendas and priorities. Scaling highlights the vital role that state actors and institutions may perform as settler ecologists at certain points in time – in this case, when state territorial interests and aspirations are in alignment with those of the national and international conservation community.

Importantly, pastoralist communities are not inherently opposed to scaling conservation. In fact, the opposite may often be true, as pastoralists, livestock, and wildlife all similarly benefit from greater ease of movement and scaled access to land. Driving between Timau and Nanyuki one afternoon in early 2023, we discussed the GBF with a young member of Il Ng'wesi. He exclaimed, "30 by 30 is not enough. We want 100 by 30! One hundred per cent of the land for people and nature!" For some pastoralists, reserving just 30 per cent of the earth for conservation seems like a small and arbitrary commitment. This individual argued that 100 per cent of the earth should be restored and protected, but in a way that would allow people, domestic animals, and plants and wild species to share space and coexist. This conviction and knowledge – that people and animals need to share space and jointly contribute to the production and maintenance of ecosystems – lies at the core of many Indigenous Peoples' ecologies in Laikipia and elsewhere around the world. It is also in line with loudening calls for convivial conservation – a Whole Earth, integrated approach to understanding and practicing environmental conservation (Büscher and Fletcher 2019). There is clearly a growing desire for more radical approaches to saving nature. Whether these approaches will end up further embedding in or departing from settler ontologies that demand separation of humans and nature remains to be seen.

Conclusion

We end the book where it began, near Sanga, a community on the northern edge of Borana Conservancy in Laikipia. We are driving up a steep road on our way out of the community. Gentle raindrops begin to fall. The sky above us is light grey and only slightly cloudy, but there are large, dark storm clouds on the horizon. We pull over the car to discuss whether we should continue up the road towards Emerua as planned for a conversation with another elder or return back to Nanyuki before the rains begin in earnest. We do not have far to go back to Nanyuki – only around 50 km – but the unpaved roads through community and private conservancies may be difficult for our small 4×4 to traverse if heavy rain falls. After such a long period of drought, the parched grounds are unable to absorb heavy rains, creating slippery driving conditions and increasing the likelihood of dangerous flash floods. A loud crack of thunder cuts through our conversation and startles a small flock of superb starlings and white-browed sparrow weavers foraging under a patch of trees beside our vehicle.

We turn to watch the birds as they settle back down in the forested patch we idle in front of, and something catches our attention. We are facing the high-voltage electric fence that separates Borana Conservancy from Il Ng'wesi Community Conservancy. Unlike other parts of the fence, this segment is surrounded by forest on both sides, except for narrow strips of land along the fence that have been cleared of vegetation to create a sightline for security reasons. What catches our attention is that the vegetation on either side of the fence looks different (see photo C.1). On Il Ng'wesi, the forest is made up of trees of different heights and colours – dark green, light green, yellow, and brown – resulting in a deep green hue overall. On Borana's side of the fence, the trees are nearly all the same height and colour – the distinctive silvery green of the African olive tree.

Our decision about the plan for the rest of the day is put on hold as we discuss the differences in vegetation on either side of the fence. From a

Photo C.1.   Varied vegetation on either side of the Borana-Il Ng'wesi fence line

distance, nearly every tree on the Borana side of the fence appears to be an African olive tree (*Olea europaea* or *Olea africana*). In contrast, from where we sit in our car on the Il Ng'wesi's side, we can see several different species: *Juniperus prospera, Acacia nilotica, Acacia Tortilis,* and *Acocanthra schemperi*. Looking beneath the trees, we note a variety of grasses and herbs as well, each adding their own colour to the forest mosaic in front of us. The discerning eye would see that, tucked among these plants, the African olive tree is also present – although more difficult to spot upon first glance, as many of the silvery green leaves have been cut away.

After some discussion, we arrive at a possible explanation for these differences in vegetation. On Borana's side of the fence, African olive trees have likely thrived at the expense of other plants. A hardy tree that is able to survive drought, hot temperatures, wind, shade, and damage, the African olive is known to form a dense shade canopy that prevents the establishment of other species' seeding and understory plants (Infonet Biovision, n.d.). In places with little human interference, the tree can grow to be large and its seeds can spread quickly with the help of birds. This is particularly true on bare land and in disturbed forests and shrublands, such as former cattle ranches like Borana. This is one of the reasons the African olive tree is at the forefront of dryland reforestation efforts.

In contrast, on Il Ng'wesi's side of the fence, African olive trees are used and managed, which allows other species to flourish. Known in Maasai as *Oloirien*, the African olive tree has many uses. The tree is used for firewood and ceremonial and medicinal purposes. The tree is also a source of fodder during the dry season and droughts. This is why most of the olive trees near us have been pruned, with only a few leafy stems left on each tree. Now that the rains have returned, regrowth is beginning and small buds are visible. Soon, these trees will be full of silvery green leaves again. In the meantime, sunlight has been able to permeate the forest canopy, reaching smaller species of trees and shrubs, while livestock grazing in the forest have reduced African olive tree seedings. Combined, pastoralists' day-to-day activities have allowed the other species mentioned above to maintain their place in the forest and continue to play their own unique roles in the larger biocultural landscape – with *Juniperus prospera* attracting bees, *Acacia tortilis* providing shade, and *Ficus thonningii* aiding in fire starting.

The forests on either side of the fence have their own unique ecological benefits and shortcomings. The forest on Il Ng'wesi's land may provide a safe haven for endemic, threatened, and culturally significant plant species that are disappearing elsewhere in the landscape (for more on the ecosystem services provided by cultural forests, see Dar et al. 2022) but provide less direct benefit to wildlife species that shy away from human activity. In contrast, the forest on Borana's side of the fence may be experiencing a decline in plant diversity, as evidenced by recent studies of the Lewa-Borana Landscape (for example, see Giesen, Giesen, and Giesen 2017), but may provide more secure refuge to growing wild animal populations. Such observations have been made in other contexts where protected and communal areas border one another, leading researchers to conclude that communal and protected areas can play complementary roles in conserving a broader spectrum of biodiversity than protected areas may achieve on their own (Nacoulma et al. 2011).

Thus, our intention in recounting these observations is not to debate whether one of the forests on either side of the fence is intrinsically better than the other. Rather, these observations offer snapshots of two different ecologies – or two different sets of relationships between humans, plants, animals, and nonliving beings, entities, and collectives (Whyte 2018b) – on either side of the fence. These different ecologies are entangled with different social structures. On one side of the fence, the ecology is a product of settler colonialism and continues to be creatively enrolled in the project of sustaining settler-colonial structures. On the other side of the fence, the ecology reflects relations that have been produced over time by pastoralists and that, in turn, shape and support pastoralism.

The sun peeks through the clouds and we decide to continue with our day as planned. The engine roars as we approach the top of the steep, rocky road and turn onto a narrow path. A troop of olive baboons sits on the path ahead, ambling out of our way as the car draws closer. We bounce along the road, heading deeper into Mukogodo Forest on our way to Emurua. We pass and greet an elderly man with about a dozen cattle heading in the same direction as us. Thorny branches slap and scrape against the car as we move through the thickening forest. A sole bushbuck crosses the path and we slow to look for a partner or calf in the dense foliage that surrounds us, but with no luck. There is no lack of biocultural diversity on Il Ng'wesi land, and yet those that live here are often told that their land does not host the right type of ecological relations. There is so much local and global pressure to extend settler ecologies into their land. From removing the high-voltage electric fence between Il Ng'wesi and Borana, to creating more secure space for rhinos to would-be lodge investors who demand a say over where pastoralists and their livestock can move in the landscape, to the introduction of carbon credit schemes that place restrictions on grazing, it is the expansionary tendencies of Laikipia's settler ecologies – that eliminate and replace as they expand – that makes them so problematic.

In this concluding chapter, we summarize key takeaway points from the book by responding to the four following questions: (1) Who are settler ecologists? (2) Where are settler ecologies located? (3) Should we study settler ecologies? And, finally, (4) what alternatives to settler ecologies exist? Our responses to these questions provide further detail and develop greater clarity around some of the book's core concepts. We also use our responses to reflect on the broader relevance of these concepts beyond Kenya. We discuss how the idea of settler ecologies could be useful in interrogating the global conservation community's response to the ongoing global biodiversity crisis. In the midst of a scramble to address unprecedented global biodiversity loss, there is a pressing need to think critically about the action being taken – particularly as new global agendas, such as GBF, endeavour to extend conservation's territorial ambitions.

Who Are Settler Ecologists?

As we explain in the introduction, settler ecologists include those who might first come to mind: whites of European ancestry, along with their descendants, who took up land with the intention of staying indefinitely in the settler colonies where they arrived. Throughout this book, we offer many examples of how white settlers have reconfigured and transformed landscapes in Laikipia over time. When a settler society was founded in Kenya over a century ago, white settlers in the region first began to enact claims to land that still stand. It is estimated that roughly one million acres of Laikipia – approximately 40 per cent of the county – is still owned by a handful of settler families (McIntosh 2017, 663; see also Letai 2015). This land provides the base on which settlers continue to re-enact their claims of belonging and advance their individual and collective interests today.

Some white settlers would contest the claim that the amount of land they hold provides them with greater influence over land and resources in Laikipia, debating their agency or ability to shape ecological relations beyond their property lines. They might also argue that the shifting political landscape since the time of Kenya's independence has made them passive landowners who have been forced to quietly adapt to a new political context in order to stay. Unarguably, the political power of whites in Kenya has changed over time. While they once held top government positions in wildlife and conservation management, this is no longer the case. There also used to be a greater number of white-dominated special interest groups, which largely dissolved in the decades after independence. Some of these special interest groups have been revitalized at different points in time with updated mandates, such as the Laikipia Farmers Association and Laikipia Wildlife Forum. Yet, many white Kenyans in Laikipia would have the opinion that their political interest and influence is limited.

Beyond having less political power than in the past, some settlers also feel their presence and belonging in Kenya is increasingly under threat (McIntosh 2017). In conversation with white settlers, it is often only a matter of time before discussions reveal anxieties about their future in the country. These anxieties are often fueled by rumours and recent events that symbolize an apparent decline of white privilege, such as white settlers being denied work permits. Ethnic violence after the 2007 election that spilled over into white properties and more recent invasions of private ranches and conservancies in Laikipia in 2017 have also heightened these anxieties, leading to debates among settlers about whether Kenya will eventually go the way of Zimbabwe, referencing the fast-track land reform program experienced there in 2000. Recognizing that settler colonialism operates through ongoing forms of dispossession that sustain access to territory – as Wolfe writes, "territoriality is settler colonialism's specific, irreducible element" (2006, 388) – threats to land tenure security rather than political control or formalized citizenship can represent a great source of anxiety for contemporary settlers.

Even though settler dominance over political systems and property regimes in Kenya may be shifting, the ability to shape ecological relations has become an alternative avenue that some white settlers use to secure their presence in the landscape. From reintroducing endangered species on their land to rescuing and breeding wild animals that require continued intervention on leased land, settlers have been able to transform ecological relations in ways that sustain their territorial claims. Moreover, as the global biodiversity crisis pushes species that dominate imaginaries of African nature closer to extinction, settlers have further secured their place in the landscape as indispensable custodians of wildlife and guardians of wild spaces. This role enables white settlers to hold on to and in some cases extend their access to territory even as their influence on national politics wanes and the amount of titled land technically under their direct ownership is in many ways diminishing.

As important as whites of European ancestry have been to influencing ecological relations in Laikipia, one of the central points we make in this book is that this group is now part of a much larger category of actors and institutions that are also engaged in creating settler ecologies. The category of settler ecologists includes a range of state and nonstate actors that all bring something different to processes of ecological transformation that sustain settler colonialism. For example, early settlers of European descent were the primary actors involved in elimination (chapter 1). However, by the time Kenya's independence era came around, not only were acts of rewilding settler land legitimized by growing concern for biodiversity loss among the international conservation community, but they were also made financially feasible by new partnerships between white Kenyans and international conservation organizations (chapter 2). Kenyan-based conservation organizations, eco-tourism ventures, public and private security actors, and even tourists are largely responsible for repeopling (chapter 3). The role of researchers and next-generation white Kenyans in settler ecologies is underscored by rescuing (chapter 4). Finally, scaling highlights just how central national environmental and wildlife authorities have become to securing a future for settler ecologies, as well as epistemic communities and zoological societies conducting research on the ecological connectivity needs of wildlife and international donors and investors providing the finances needed for scaling (chapter 5).

When we speak of settler ecologists, we refer to all those who consciously and unconsciously participate in the elimination and replacement of existing ecological relations with those that enable structures of settler colonialism to endure. Settler ecologists, therefore, are not defined strictly by their ancestry or race, but instead by their contributions to destructively creating ecological relations that prolong settler colonialism as a structure. The broad definition of settler ecologists that we adopt captures the wider assemblage of actors involved in creating and sustaining settler ecologies and enables us to avoid speaking of settler colonialism as an anachronism. Now more than ever before, this wider cast of characters plays an essential role in providing the finance, expertise, and authority needed to bring settler ecologies to life. These actors work in tandem to (re)produce species and interactions between species that are seen as essential to addressing biodiversity loss and, consequently, help to shore up access to territory.

This is not to say that all these characters hold the same degrees of power, agency, or conviction in their roles as settler ecologists. For example, those investing in, establishing, and managing eco-tourism ventures that mandate that certain types of ecological relations become (in)visible are far more responsible for the reproduction of settler ecologies than the labourers that work for these ventures. Although these labourers – the rangers, the guides, the security guards, the gardeners, and the maintenance people – provide the physical work needed to sustain settler ecologies, their agency in these roles may be constrained if not altogether exploited (Coe and Jordhus-Lier 2011; Thakholi 2021). In this sense, both history and context shape how various actors are differentially enrolled in and responsible for the creation of settler ecologies.

Where Are Settler Ecologies Located?

The analysis presented in this book is historically grounded and place based, focusing on Laikipia, Kenya. Laikipia is widely acclaimed in global conservation circles for its successes in biodiversity conservation. However, work by Mbaria and Ogada (2016) also shows how this success is based on and preserves colonial land ownership structures and other forms of coloniality, with conservation actors able to build lodges, erect fences, and conduct research on private land with relatively minimal involvement from the government compared to similar operations in national parks. Building on this work, we show how settler colonialism has also transformed the ecology of place in Laikipia, reconfiguring and replacing ecologies to create those that secure past settler-colonial advances. In doing so, we offer unique insights into how settler colonialism persists as a structure in Kenya.

Kenya is somewhat distinct as a settler colony, as white settlers have always represented a minority of the population. Yet, despite being a relatively small and inward-looking community with marginal influence in national politics today, white Kenyans have been effective at using their connections and networks outside the country to bolster their influence. Many early settlers to Kenya were aristocrats with close connections to British elites. These types of connections can still be observed – for example, in the international social networks they maintain with peerage, royalty, and members of the British political class and their frequent travel to and from boarding schools, universities, family estates, and so on in the UK.

Over time, these external connections have shifted and expanded towards the conservation sector, with many white Kenyans now deriving power, influence, and legitimacy through external conservation actors, including international conservation organizations, conservation researchers, conservation news media like *National Geographic*, conservation investors, and foreign tourists. Even the British Army Training Unit, which maintains a base in Nanyuki on land seized from Maasais, now works to legitimize its continued presence in the landscape through its involvement in conservation projects and training Kenya Defence Forces in combating poaching. Conservation has allowed white Kenyans in Laikipia to reframe themselves as bastions of biodiversity. This places them in a uniquely powerful position to shape what the future of the conservation landscape looks like in Laikipia, elsewhere in Kenya, and eastern and southern Africa too, with Laikipia often referred to as *the* model for private conservation. Our focus on Laikipia reveals how settler-colonial structures in East Africa are enlivened and maintained today through transnational circuits of influence, including those in the conservation sector.

Our empirical focus on Kenya is significant, as there is a geographic bias in existing literature on settler-colonial contexts in the global North and so-called Western world (see Mar and Edmonds 2010; Peake and Sheppard 2014; Pasternak and Dafnos 2018). Although these contexts provide a strong empirical base for the subfield of settler colonial studies, settler-colonial contexts outside of Australasia, North America, and Israel/Palestine have yet to receive the full scholarly attention they deserve. this means that understanding of settler colonialism's continuities and discontinuities globally remains limited and partial, as it is based largely on contexts where settlers make up a majority of the population. The flipside of this coin is that knowledge of settler colonialism in contexts like eastern and southern Africa tends to be based on historical analyses, positioning settler colonialism as a structure of the past rather than the present in postcolonial Africa. By focusing on the settler-colonial present in Kenya, this book contributes to growing the geographical and temporal scope of scholarship on places settlers came to stay – expanding the geography of settler colonial studies.

At the same time, despite the importance and distinctiveness of focusing on Kenya, we do not see settler ecologies as unique to Kenya. Building on relevant work in environmental history (Crosby 1989; Griffiths and Robin 1997; Beinart 2003; Beinart and Hughes 2007; Gissibl 2016), we argue that the constant unmaking and remaking of ecologies on expropriated land is a defining feature of how settler colonialism operates as a structure and endures today. Referring to land regimes and governance systems, Wolfe writes that "settler colonialism destroys to replace" (Wolfe 2006, 388). In *Settler Ecologies*, we show how this observation applies to ecological relations in settler colonies too. The species and ecological relations that become the focus of settler ecologists may vary from one geographical context to the next. Yet the use of ecological relations to prop up and prolong settler colonialism is likely to take place wherever settler colonialism persists, as settler colonialism is a more-than-human and multispecies endeavour.

Some of the approaches to ecological transformation we identify and explore in this book – eliminating, rewilding, repeopling, rescuing, and scaling – likely have currency and application beyond Kenya. Parts of South Africa, Zimbabwe, Botswana, and Namibia, where settlers arrived "with an intention to stay permanently" (Velednitsky, Hughes, and Machold 2020, 3), have undergone similar ecological transformations to those we discuss in Kenya. Across these countries, settlers set about eliminating ecological relations that stood in the way of establishing a colonial political economy of production by hunting wild species prone to destroying crops and preying on livestock and supplementing their incomes through safaris and trophy hunting. Settler ecologists in these contexts also later went about rescuing many of the same species their hunting practices had only recently decimated, such as lions and rhinos, in fairly similar ways to what we outline above, often relying on enclosing and fortifying land to create safe havens for wildlife (Suzuki 2001; Kamuti 2014; Heydinger 2021). Contextual differences undoubtedly exist in how these ecologies were unmade and remade; yet, across the board, a preoccupation with similar species of mammals at similar points in time can be observed.

In other settler colonies, the approaches to ecological transformation we outline above may serve as a point of departure for zooming in on how ecological relations were changed by colonial settlement and continue to be reflected in biological diversity and ecological relations today. For example, Canada and the United States experienced a period of widespread ecological devastation in the centuries following colonial invasion. As Kyle Whyte (2018) describes, US settler-colonial domination of the Anishinaabewaki worked to eliminate existing ecologies so that new ecologies could be destructively created, as settler mining, deforestation, and industrialization in the Great Lakes regions killed off and decimated species at the same time that new species were unintentionally and intentionally introduced. These actions of ecocide resemble what was being done in Kenya around the time of colonial settlement.

However, there are examples of cases where ecological transformation has started to unfold in North America somewhat differently than in Kenya. For example, recent attempts to reintroduce species subjected to elimination are being undertaken collaboratively by government organizations and First Nations in ways that place greater emphasis on cultural-ecological reconnection and restoration (Taschereau Mamers 2020). Though likely still shaped by certain legacies of settler colonialism, these efforts appear to do a better job of centring Indigenous-led governance and developing meaningful co-management arrangements compared to what may be seen in Laikipia, for example (Artelle et al. 2021). This highlights the variation that exists across space in the ecological transformation being driven by or in response to settler colonialism.

Should We Study Settler Ecologies?

Earlier, we mentioned several points of discomfort we experienced in writing this book. One of these points relates to the question of whether we, and others, should direct attention to settler ecologies in the first place. As we state in the introduction, our analytical focus comes with the risk of re-centring settler ecologies in stories about the past, present, and future of Laikipia, while relegating ecologies otherwise to peripheral roles on the sidelines. This is a concern with many studies on privileged social groups and structured power relations, which may be critiqued for further normalizing, rather than dismantling, racist structures. Some might argue that focusing on settler ecologies risks reproducing the very same "logic of elimination" (Wolfe 2006) that it aims to critique by upholding the problematic notion that settler ecologies are the only ones that matter or continue to exist in settler-colonial contexts. As Fung warns, settler colonial studies can quite easily become "in itself a form of white supremacy by perpetuating the same systemic erasures of anti-Blackness, minimising racialised stories, and decentering Indigenous knowledge" (2021, 115).

And yet, we were taught from our earliest formative years that the exclusionary spaces we went to during school holidays or when friends visited from abroad – which were often operated by old settler families or more recent whites settling on the continent – were essential to the "salvation of African nature" (Nelson 2003). Without these spaces, the animals that have now been made to epitomize African nature would disappear forever, so the story would go. There would be no more elephants lazily moving through the bush at dusk or lions calling in the early hours of the morning; no more kudus, leopards, rhinos, or zebras that we loved spending hours searching for or watching at watering holes. We grew up visiting the very wildlife sanctuaries we now write about, eagerly interacting with animals being fed and cared for, possibly to be released back into the wild one day, where we listened with admiration and fascination about the work being done to rescue species at the brink of extinction.

In such spaces, problematic narratives circulated about the heroes, victims, and villains of biodiversity. All too often, white conservationists were framed as saving Africa's helpless wildlife from racialized local communities engaged in poaching, charcoal production, and other activities deemed to be unsustainable. Although the hue of these narratives has somewhat shifted over time, with increasing emphasis being placed on the important role of black and Indigenous Kenyans in conservation, the "whiteness of power" (Goudge 2003) in conservation remains an unresolved issue. By this we mean that often unacknowledged assumptions of white, Western superiority shape the discourses and practices of conservation globally. In Laikipia, and likely many other settler-colonial settings, the implication of this is that settler ecologies are often upheld as the only hope for saving biodiversity.

Thus, without downplaying or claiming to fully evade the risk of re-centring settler ecologies, it remains our conviction that there is a need to name and make visible the way ecological relations can be organized to draw settler-colonial violence into the present. By bringing attention to the assumptions, biases, and power relations that inform how and why ecological relations are organized, we neither normalize nor uphold these ecologies. Instead, we render the familiar and romanticized species and landscapes of settler ecologies strange and open them up for debate. Informed by Zoe Todd's thinking, we see "understanding how settler colonialism structures itself in the lands, waters, and atmospheres that it invades [as giving] us the power to refract its efforts and assert something liberatory in its place" (2017, para. 3).

Multispecies perspectives of settler colonialism are therefore essential to "revealing, supporting and pursuing radically different ecologies that are more inclusive and just" (Mabele et al. 2021). By revealing how relations between humans, animals, plants, and other entities in landscapes can be seized and altered to serve settler-colonial interests and prop up structures of settler colonialism, we hope this book is of use in the pursuit of other ecologies, including those that once existed, that persist in defiance of settler colonialism, or that could exist in the future.

What Alternatives to Settler Ecologies Exist?

There is a recurring question we have been asked when speaking about this work over the last several years: What alternatives to settler ecologies exist? In some cases, this question has been asked with anticipation, the person asking it appearing keen to learn of a never-settled or recently unsettled ecology in Laikipia that stands in defiance of settler colonialism. In other cases, this question has been asked as a challenge, with the person posing the question unconvinced that the particular set of ecological arrangements we speak about in this book are really a result of settler colonialism and demanding a counterfactual. Our answer to this question has often been complicated, messy, and, likely to some, insufficient. Here, we explain why this is not a question we engage with at length in this book.

To begin, we have been hesitant to make claims about knowing ecologies otherwise or prescriptions about how ecologies otherwise should look. As we state in the introduction, our positionality makes us well suited to study and speak critically of settler ecologies – to "refuse, refract, and resist" (Todd 2017) settler colonialism in its many forms, including how it manifests in and through ecological relations. At the same time, our positionality places constraints on the contributions we feel we can and should make. From our position as white settler academics, our understandings of land and life "have been shaped in powerful ways by Euro-Western conceptual frameworks and critiques circulating within such frames" (Taschereau Mamers 2020, 128). Although there may come a time when we speak alongside others about how ecologies otherwise might be (re)created and sustained, for now we remain primarily committed to listening to and learning from non-settlers about their ecologies. In other words, we do not speak at length about ecologies otherwise as we acknowledge where our own expertise begins and ends and understand that there are others who are far better positioned than ourselves to define ecologies otherwise.

At the same time, it is important that our relative silence on this topic is not misinterpreted as meaning that ecologies otherwise do not exist or do not matter. The elimination of Indigenous ecologies may form part of the ontological core of settler colonialism; nevertheless, indigeneity "exists, resists, and persists" (Kauanui 2017, 1). Although settler-colonial regimes retain much power, they are ultimately failed projects in the sense that they proved unable to rid colonies of Indigenous populations as they intended to (Joudah 2020). Furthermore, just as indigeneity itself has endured colonial violence (Kauanui 2017), indigenous ecologies have endured too. Although this book reveals and explains the practices and processes of elimination carried out by settlers and sustained through settler ecologies, we acknowledge that other ecologies were not entirely eliminated. In fact, quite the opposite is true.

In Laikipia, large parts of the landscape continue to be produced and maintained by Indigenous pastoralists, including groups of Maasais, Samburus, Turkanas, and others in the broader landscape. Pastoralists continue to reassert their ecological practices through their livelihoods in a landscape where they have been squeezed out and suppressed for well over a century. Their use of the environment – including animals, herbs, and trees and water – along with their livestock's grazing patterns, directions, and timings of movements, continues to influence the composition and distribution of plant and animal species in the landscape. Despite the efforts of the colonial government to stamp out their ways of life, the ecological relations that pastoralists and their livestock create are imprinted on the land.

It is well known that pastoralist ecologies often involve complementary and codependent relationships between pastoralists, livestock, and wild animal and plant species, and that these have taken shape over the years through highly adaptive and flexible systems of communal land use. Well-managed and moderately stocked herds help to create nutrient-rich patches of land that contribute to plant diversity and provide palatable forage for animals. This promotes rangeland health by improving soil fertility, supporting biodiversity, and lessening fire risk. At the same time, moving herds about the landscape allows livestock to forage across environments that vary greatly in altitude, moisture, and vegetation type and achieve a balance of different nutrients in their diet. This movement also enables land to recover and renew fairly rapidly following rainfall (Melubo 2020). With this in mind, the ecologies many pastoralists value and create have the potential to address many of the challenges settler ecologists have faced in the region over the last century – including environmental degradation and disease or poor health – in trying to sustain large populations of cattle and, later, wildlife on fenced land. These ecologies provide an example of alternatives that exist to settler ecologies.

Although pastoralists' ecologies are certainly one example of ecologies otherwise, they are not the only example.[1] A risk of defining ecologies otherwise strictly as pastoralist ecologies is that of fixing pastoralists

---

1 In line with this view of ecologies otherwise, Goldman's (2022) work offers valuable insights into how the storytelling practices of Maasai groups across eastern Africa can serve to decentre conventional scientific knowledge of nature and environmental conservation.

to the landscape in particular ways and locking them into very specific types of environmental roles and relations. As Tania Li writes, there is a trend where Indigenous Peoples' rights are recognized by contemporary conservation initiatives, but only if people remain fixed in prior and unchanging relationships with nature (Li 2014). This is a further reason we have been reluctant to define ecologies otherwise in this book: we want to avoid painting ecologies otherwise in Laikipia as homogeneous, predetermined, or static.

For example, during a conversation with Earnest, an Indigenous rights activist from Il Ng'wesi who has recently been involved in advising intergovernmental panels about Indigenous rights and global biodiversity conservation finance, we spoke about Wildlife Conservation Bonds. These bonds provide a conservation success payment to shareholders that is dependent on verified increases in wildlife population numbers. Earnest explained how Wildlife Conservation Bond schemes could be redesigned to recognize the value of pre-existing Indigenous systems for biodiversity conservation. Each Maasai family has a totemic affiliation with different animal species; one family may recognize rhinos as totemic or sacred and have a responsibility to protect rhinos as a result while another family may have the same responsibility to elephants. Historically, the sacred value of these species helped protect wildlife – especially when an animal inflicted some form of harm on people – as no action was permitted against any totemic animal without consent from its corresponding human family. Earnest suggested Wildlife Conservation Bonds could be used to acknowledge the conservation value of the Maasai totem system just as other conservation actors in the landscape are financially supported for their contributions to conservation. Some critics of neoliberal conservation have raised concerns about how mechanisms like Wildlife Conservation Bonds risk assigning monetary value to wildlife protection in ways that reorganize and rewrite peoples' relationships with nature (Fletcher 2023). However, in Earnest's view, this need not always be the case with Wildlife Conservation Bonds. Instead, he described how these bonds could represent an opportunity to reinforce relations between pastoralists and wildlife that are guided by Indigenous, rather than settler-colonial, logic – such as those upheld through the totem system.

There is no straightforward answer or singular definition to questions about how ecologies otherwise should be defined in Laikipia. A century ago, it may have been possible to point towards certain species or a specific set of ecological relations that could be understood more strictly as Indigenous – providing a clear counterpoint to settler ecologies. Still today, there are likely some species that are more important to ecologies otherwise, such as livestock or certain types of grasses, herbs, and trees. Yet, the arrival of European settlers over a century ago set in place a slow-motion ecological explosion and protracted upheaval of many Indigenous ecological relations. Other changes have also since taken place. From the spread of globalization and global capitalism to unprecedented environmental and land use change, Laikipia's ecologies otherwise remain in a state of flux and ongoing negotiation. Therefore, today, it is not the absence or presence of particular plant and animal species that differentiates settler ecologies from ecologies otherwise. In the simplest of terms, ecologies otherwise are those that most clearly sustain Indigenous Peoples' lifeworlds and refract and resist settler-colonial power.

Conclusion

Settler Ecologies is not just an explanatory story about the past and present of settler colonialism in Kenya; it is also a cautionary tale for the future of global conservation action. With growing concern about global biodiversity loss and threat of a mass extinction event looming, a global conservation scramble is underway. This scramble is being made possible by new partnerships and novel sources of finance that aim to free up more space for conservation. This could facilitate new opportunities to forge pathways to ecologies otherwise. However, as we discuss in the latter chapters of this book, many recent responses to the global biodiversity crisis risk reproducing settler ecologies in settler-colonial contexts and expanding settler ecologies to other contexts and new frontiers of conservation instead.

The 2020s are poised to be crucial for biodiversity, following the IUCN World Conservation Congress in Marseille, France, and the newly adopted global biodiversity agreement, the Kunming-Montreal Global Biodiversity Framework (GBF), of 2022. A total of 188 governments signed onto the GBF, agreeing to address the ongoing loss of terrestrial and marine biodiversity. With these commitments, massive amounts of funding are now being made available to support the implementation of the GBF, with signatories working towards raising US$200 billion per year in domestic and international biodiversity-related funding from public and private sources.

A focal point of the GBF is substantially increasing the total area of natural ecosystems under protection globally. The agreed-upon target is to protect 30 per cent of the planet by 2030; known informally as "30 by 30." Currently, efforts to achieve 30 by 30 emphasize the expansion and creation of national parks as well as the establishment of OECMs, including wildlife corridors, dispersal areas, and private or communal conservation areas. The expansionary ambitions of the GBF have concerned many human and Indigenous rights organizations and activists, due to a worrying lack of engagement with the historical injustices and colonial legacies of protected areas during the drafting of the GBF (Cariño and Ferrari 2021; Reyes-García et al. 2022; Loring and Moola 2021). In light of the well-documented, problematic history between Indigenous Peoples and protected areas, a fundamental question about the GBF remains unanswered: Whose ecologies are going to be represented in lands and territories being acquired for 30 by 30?

In preparing the groundwork for the global territorial expansion of conservation, the GBF makes no mention of the imperative to re-establish the rights of Indigenous Peoples where they have been extinguished through conservation action (Rainforest Foundation UK and Survival 2020), nor does it acknowledge the importance of securing land tenure for Indigenous Peoples before land is demarcated and acquired for conservation. On top of this, currently only "1% of the funds available for climate and the environment goes to IPLCs [Indigenous Peoples and local communities]" (Gjefsen 2021). This means that, in comparison to the billions of dollars of conservation funding tied to the GBF, only a miniscule amount is being directed at IPLCs and ecologies otherwise. This reality is fueling concerns that the post-2020 era of biodiversity conservation will continue to bear the hallmarks of colonial and fortress conservation models that dominated most of the previous century. The potential justice implications of these trends for IPLCs are increasingly apparent; but it remains under-appreciated that injustice against IPLCs translates into missed opportunities for global biological diversity and ecosystem health as well.

A growing number of studies support what IPLCs have long known and proven: there are positive outcomes for biocultural diversity and ecosystem health when IPLCs play a leading role in conservation (for example, see Sze et al. 2022). In the context of hegemonic settler ecologies, creating the conditions that would allow IPLCs to pursue and realize their own ecologies requires a dramatic shift in conservation finance and action. Resources need to be channeled directly to Indigenous Peoples to unmake settler ecologies and remake ecological relations that better reflect their evolving aspirations and interests. This shift needs to be matched by changes to policy and legislative frameworks that give IPLCs the freedom and autonomy needed to undertake reparative, restorative forms of ecological transformation. The unmaking and remaking of settler – and other unjust – ecologies may not initially unfold in ways that yield the typical conservation outcomes and timelines required by donors, investors, and authorities. Yet what may at first look like "a programme of complete disorder," as Frantz Fanon (1963, 36) writes, may still be necessary if a plurality of ecologies is to take root around the world and give seed to radically different and more abundant and truly (bio)diverse futures.

Even in the absence of just policy and legislative frameworks, as well as adequate global support, individuals and grassroots organizations in Laikipia are attempting to undo the ecological violence caused by settler colonialism. They are supporting pastoralists communities in securing communal land tenure, removing invasive species, maintaining the health of their livestock, halting the degradation of pasture and soil, and (re)planting grasses, herbs, and trees that support pastoralist ecologies. In the past, funding for these projects was often secured in atypical ways, including from private trusts or epistemic institutions. However, it is encouraging to note that some well-known conservation organizations are also beginning to fund this work. For example, the Global Environment Facility and Conservation International have recently granted funds to one organization coordinating the establishment of ICCAs (Indigenous and Community Conserved Areas) in Laikipia through the Inclusive Conservation Initiative. Although this initiative was in its infancy when this book was published, and was by no means problem-free or certain in its outcomes, it is symbolic of the type of shift in conservation finance and action needed to begin undoing settler ecologies and pursuing ecologies otherwise.

In May 2023, we visited a small owner-operated lodge in the Waso area of Laikipia to attend meetings for the Inclusive Conservation Initiative. The owner of the lodge, who is from the community, has been working to restore small-scale versions of ecologies otherwise around the premises. Only a traditional-style *boma* and, in some places single wire, demarcate the lodge grounds, allowing animals, plants, and people to come and go with relative ease. Outside the *boma*, a slow-motion ecological struggle has unfolded, creating a landscape that looks far more uniform than it would have historically – now dominated by *Opuntia*, *Euphorbia candelabrum*, and *Sansevieria*. Yet, on the lodge grounds, the owner has been removing *Opuntia* plants one by one, along with other plants of less value to pastoralism, and replacing them with indigenous grasses, herbs, and shrubs and trees. The grounds now display many young trees and herbs and shrubs, including whistling thorn acacia (*Vachellia drepanolobium*), mgunga (*Faidherbia albida*), prickly thorn (*Senegalia brevispica*), and white crossberry (*Grewia tenax*). These surround the stone-lined paths and open-air, thatched-roof bandas strategically spread across the property.

One morning at the lodge, we sit under a large *Ltepes* tree, next to a fire pit still smouldering from the previous night, watching elephants stride across an adjacent hill on their way to the Ngare Narok for water. Two elders from the nearest community enter the main hall of the lodge and walk towards us carrying short wooden stools. As they plunk their stools down in the ankle-high grass sprouting up beneath the tree, we exchange greetings: "*Sopa*." One of the elders spits and closes his eyes, taking a long, deep breath in through his nose. The other shields his eyes with a hand and surveils the grounds as a smile forms on his face. "This place is beautiful. See all our herbs and trees?" he asks us, motioning with his other hand. The pride and rejuvenation evident in the elders' body language and words strike us. Even in the midst of powerful and enduring settler ecologies – in the interstitial spaces between settler-owned conservancies – reparative and restorative ecological transformations are taking place. As these root, bloom, and seed after recent rains, they keep the promise of ecologies otherwise alive.

Afterword

This afterword is a continuation of the foreword written by Ramson Karmushu, describing a vision for the future in which settler ecologies are no longer domi- nant on the Laikipia Plateau and its surrounding lowlands and highlands.

Fences do not simply divide land. They separate people and nature. I do not believe in this separation, nor do many in my community. Maa- sais believe in nature: "we are a living nature," as we always say; "we are nature and nature is us." We believe in coexisting with nature. It is not by chance that we are in nature; it is one of our basic needs. Fences are a barrier to coexistence. It is like creating hostility with nature. We need nature and nature needs us. Separating us from nature is injustice to nature and, in my eyes, unsustainable.

It is difficult for me to imagine a future without fences, as conserva- tionists have long been busy fundraising for this indefensible ideology. Truthfully, there are times these days when I wish to fence a plot of land for myself and my family, especially when I think about initiatives like Ukanda wa Vifaru, which proposes to remove the fence between Borana, Lewa, and Il Ng'wesi and construct a fence around the remain- ing perimeter of Il Ng'wesi – creating a land contiguous with the Lewa- Borana Landscape. I once had the misfortune of talking with a conser- vationist, who was asking me about this Il Ng'wesi fencing proposal. Little did he know that at the time, in October 2018, neither I nor the Il Ng'wesi community had clear information on why changes to the Il Ng'wesi fence were being pushed by Lewa and Borana – the "kingpins" of conservation where I am from. I insisted to this conservationist that only the community could decide what would happen to Il Ng'wesi's fence. He seemed very unhappy with my response and exclaimed, "it should happen and very fast!" The conservationist continued to say that the land belongs to rhinos and, in the long run, the community would be removed. I term this as new colonization that is being sang by the current leaders of the boards in conservation who are serving their own interests. I was so astonished by this response. It made me pause and ask myself, "is it true or real that someone seated in London knows Il Ng'wesi better than I do?" It made me question my community land membership and the procedure any fence removal and construction would employ, and whether it would respect the free prior and informed consent (FPIC) of Indigenous Peoples. Funnily enough, this discussion was not in Il Ng'wesi, or even in Kenya; yet someone seated in an office in London felt they knew better what should happen in my community's land – my land.

I had always been keen to learn about the Indigenous and Traditional Knowledge (ITK) different communities in northern Kenya used to live in harmony with nature and support their own conservation practices. However, this conversation gave me new motivation to better understand the traditional knowledge and ways of my own people and to advocate for inclusive, sustainable approaches to conservation that centre these ways. It also motivated me to look into local, regional, and international frameworks, such as the Aichi Targets, Nagoya Protocol, and Convention on Biological Diversity, as well as local conservation agendas that support or affect coexistence. After London, I took a great interest in designing research initiatives, like Decolonising Human-Wildlife Conflict (HWC), which sought to document ITK concerned with preventing HWC and promote harmonious relationships with nature. Through Decolonising HWC, I also produced an Indigenous field guide to Mukogodo Forest, which is located within Il Ng'wesi. I am grateful to Charis and Brock for supporting me in different ways to realize such dreams.

The knowledge documented through such research initiatives is incredible, but eroding at high speeds, as it is written in community libraries (elders' heads) – and more of these disappear with each passing year. This knowledge can be regenerated through community-friendly conservation projects, such as the Inclusive Conservation Initiative, which I believe is sustainable and being driven by members of Indigenous communities.

Conservation practitioners, policymakers, and scientists need to respect Indigenous Peoples, who have been living *with* nature for centuries and who hold the keys to a biodiverse and healthy future for the planet. This future cannot be achieved through more fences and fortress conservation models. Conservationists need to take pride in the knowledge and ways of Indigenous Peoples and support their existing capacity for inclusive, sustainable conservation. As one elderly man from my community once asked me, why are the white people telling us how to conserve wild animals while we hear people say there are no wild animals in their lands? They should come sit and listen to us and we will tell them how to live respectfully with wildlife and nature.

With this in mind, when I try to picture a future without fences, I always try to hear the voices of the elders in my head. One of the first things that comes to mind is wildlife corridors. Without fences, there would be no need to construct fence gaps or to fence off corridors; this is colonizing nature and wildlife. It should be noted that most of the so-called wildlife corridors in Laikipia today have historically been wildlife *and* livestock corridors. I always wonder why conservationists talk of wildlife corridors without livestock. This, to me, is just creating enmity and advancing human rights violations by criminalizing people and their livestock in their own lands. There should be only migratory routes across an open landscape, maintained by livestock and wild animals, such as buffalos and elephants. Pastoralists have now been forced to migrate using roads like cars because our migratory routes have been named wildlife corridors and grabbed inside the settler ranches or conservancies. As in the past, these corridors would be dual usage. They would also allow people and livestock to migrate from the highlands to the lowlands and back again – movements now restricted by fences and security teams.

To imagine a future without fences, I simply imagine the past. However, there are also new technologies and tools available that can assist us in coexisting with wildlife and ensuring the protection of wildlife. These days, there are many financing arrangements for conservation, although most of these are not shared with my community – "Rhino Bonds," for example. These sources of financing can be made to work for Maasais in ways that are more beneficial and empowering, which is why conservationists never share them directly with us. They only mediate between communities and donors and investors.

I can imagine new financing arrangements for rhino conservation in communal lands managed by pastoralists. We could have an arrangement whereby anybody who wants to harvest rhino horns pays individual families or communities to guard and care for one rhino – the rhino can even be given the name of the investor or a name of the investor's choosing. After so many years – or after a rhino has produced however many offspring – the sponsor would be free to harvest the horn with our blessing. This could allow for the protection of rhinos without the need for fences that fragment our landscapes and impede access to nature. Furthermore, although we do not understand why people purchase such products, if someone cannot do without, special arrangements can be made and the products could be harvested without killing the animal.

Without fences, wildlife might begin to stay with us the way it did in the past. The behaviour of animals might begin to change back to normal. In chapter 5, Charis and Brock discuss some findings from research we carried out together in Il Ng'wesi. During this research, elders in Il Ng'wesi explained how conservancy fences are changing elephant behaviour, forcing elephants to stay longer in our communities, where they become anxious and arrogant and behave inappropriately. Gaps in fences can *never* fully address the problem of landscape fragmentation created by the same fences. The same elders also believed that there has been a great change in animal behaviour that is contributed to by the lodges in the conservancies who provide them with salts and water. This is not sustainable because when the grass is depleted in the conservancy areas, animals have to move to community lands. Since we do not provide salt for elephants in our settlements, elephants now forcefully demand it from our houses and livestock troughs, raising conflicts.

Conservationists often blame conflict with wild animals – especially elephants – on our people, saying elephants fear us and therefore act aggressively. These stories deny Maasais our history. For centuries, pastoralists have lived together with wildlife and we can prove this, as there are spaces in northern Kenya's rangelands without conservancies where wild animals can still be found alongside pastoralists and livestock. The Maasai have always managed wildlife so well. The elders also believe that when an elephant kills a human and is not eliminated, it continues to kill and disturb people. According to traditional ways of knowing, the animal that killed a person needs to be immediately killed in turn. This practice still happens on settler ranches in Kenya, as well as with other animals in Western countries, like Canada; however, my people continue to be denied this practice on our very own land – where we are custodians and guardians of nature.

The lifestyles and traditional beliefs of many pastoralist communities also prove we can still coexist. For example, many Maasai families have their own totems. Some have the hyena, some have the elephant, some have the lion, and so on. Before conservation, when these totems had a function, each family was responsible for a different totem. If an elephant needed to be punished, the responsible family was the only one who could allow and give permission for this to happen; respecting it like a brother, they would not participate in killing the animal. So the management of wildlife was clearer right down to the level of individuals. These beliefs and practices helped us maintain healthy populations of wildlife by only eliminating problem or rogue individuals.

I also believe that if we had the whole of this blended landscape, there would be much greater knowledge of how to avoid conflict with wildlife across our people. We would be able to revitalize our knowledge. When we used to interact daily with all animals, elders devoted time to training children on how to escape, avoid, and survive conflict situations with wildlife. For example, growing up, I was very clearly shown how to avoid an elephant fight and how to avoid a rhino. For a buffalo, you have to climb a tree; otherwise, lie down on the ground if there are no trees and they will lick you with their rough tongue instead of using their horns. Whenever you lie down, use something sharp to knock on their nose and they will definitely run away. Or, when you want to run and escape a buffalo, you run uphill. If you want to escape an elephant, you run downhill.

If we had a really good landscape where there were no fences and we moved freely, if our landscape was blended for use between wildlife and livestock and people, and if we had enough land, the meanings and practices that enabled us to live with nature in the past would be revitalized. I want to tell the settlers on Ilaikipia soil, our ancestors' soils, "Keep running; you will eventually get tired, and your ways will diminish. I want to reassure you that, even if we are only crawling now, we will catch up."

– Ramson Karmushu, Nanyuki, Kenya, May 2023

References

&Beyond. n.d. "Sarara Camp." Samburu Wilderness. Accessed 13 October 2023. https://www.andbeyond.com/places-to-stay/africa/kenya/samburu -national-park/sarara-camp/.

Adams, William Bill, and Martin Mulligan. 2012. *Decolonizing Nature: Strategies for Conservation in a Post-colonial Era*. London: Routledge.

Adams, William M. 2009. "Sportsman's Shot, Poacher's Pot: Hunting, Local People and the History of Conservation." In *Recreational Hunting, Conservation and Rural Livelihoods: Science and Practice*, edited by Barney Dickson, Jonathan Hutton, and William A. Adams, 127–40. Chichester: John Wiley and Sons.

– 2020. "Geographies of Conservation III: Nature's Spaces." *Progress in Human Geography* 44, no. 4 (August): 789–801. https://doi.org/10.1177/0309132519 837779.

Agutu, Nancy. 2017. "States Declares Seven Laikipia Areas 'Dangerous, Disturbed.'" *The Star*, 15 March 2017. https://www.the-star.co.ke /news/2017/03/15/states-declares-seven-laikipia-areas-dangerous -disturbed_c1525640.

Akama, John S., Christopher L. Lant, and G. Wesley Burnett. 1996. "A Political-Ecology Approach to Wildlife Conservation in Kenya." *Environmental Values* 5, no. 4 (November): 335–47. https://doi.org/10.3197/096327196776679276.

Akama, John S., Shem Maingi, and Blanca A. Camargo. 2011. "Wildlife Conservation, Safari Tourism and the Role of Tourism Certification in Kenya: A Postcolonial Critique." *Tourism Recreation Research* 36 (3): 281–91. https://doi.org/10.1080/02508281.2011.11081673.

Anderson, David. 1984. "Depression, Dust Bowl, Demography, and Drought: The Colonial State and Soil Conservation in East Africa during the 1930s." *African Affairs* 83, no. 332 (July): 321–43.

– 2005. "'Yours in Struggle for Majimbo': Nationalism and the Party Politics of Decolonization in Kenya, 1955–64." *Journal of Contemporary History* 40, no. 3 (July): 547–64.

Anderson, David, and David Throup. 1985. "Africans and Agricultural Production in Colonial Kenya: The Myth of the War as a Watershed." *Journal of African History* 26, no. 4 (October): 327–45. https://doi.org/10.1017/s0021853700028772.

Andrewartha, Herbert George, and L. Charles Birch. 1954. *The Distribution and Abundance of Animals.* Chicago: University of Chicago Press.

Anonymous Reviewer (@BJ0707). 2013. "This place is pure magic…" Review of Il Ngwesi Lodge on TripAdvisor. 31 October 2013. https://www.tripadvisor.co.uk/ShowUserReviews-g612348-d594072-r183094670-Il_Ngwesi_Lodge-Nanyuki_Town_Nanyuki_Municipality_Laikipia_County_Rift_Valley_Pro.html.

Artelle, K.A., M.S. Adams, H.M. Bryan, C.T. Darimont, J. Housty, W.G. (Dúqváísḷa) Housty, J.E. Moody, et al. 2021. "Decolonial Model of Environmental Management and Conservation: Insights from Indigenous-Led Grizzly Bear Stewardship in the Great Bear Rainforest." *Ethics, Policy & Environment* 24, no. 3 (September 2021): 283–323. https://doi.org/10.1080/21550085.2021.2002624.

Aviation, Travel, and Conservation (ACT) News. 2018. "Rhino Ark Charitable Trust and Rhino Charge – A Look Behind the Scenes." *ATC News*, 1 June 2018. https://atcnews.org/rhino-ark-charitable-trust-and-rhino-charge-a-look-behind-the-scenes/.

Ávila-García, Patricia, and Eduardo Luna Sánchez. 2012. "The Environmentalism of the Rich and the Privatization of Nature: High-End Tourism on the Mexican Coast." *Latin American Perspectives* 39, no. 6 (November): 51–67. https://doi.org/10.1177/0094582x12459329.

Bates, Robert H. 1987. "The Agrarian Origins of Mau Mau: A Structural Account." *Agricultural History* 61, no. 1 (Winter): 1–28.

Beinart, William. 2003. *The Rise of Conservation in South Africa: Settlers, Livestock, and the Environment 1770–1950.* Oxford: Oxford University Press.

Beinart, William, and Lotte Hughes. 2007. *Environment and Empire.* Oxford: Oxford University Press.

Benjaminsen, Tor A., and Ian Bryceson. 2012. "Conservation, Green/Blue Grabbing and Accumulation by Dispossession in Tanzania." *Journal of Peasant Studies* 39, no. 2 (April): 335–55. https://doi.org/10.1080/03066150.2012.667405.

Benton, Adia. 2016. "African Expatriates and Race in the Anthropology of Humanitarianism." *Critical African Studies* 8, no. 3 (September): 266–77. https://doi.org/10.1080/21681392.2016.1244956.

Bersaglio, Brock, and Jared Margulies. 2022. "Extinctionscapes: Spatializing the Commodification of Animal Lives and Afterlives in Conservation Landscapes." *Social & Cultural Geography* 23, no. 1 (January): 10–28. https://doi.org/10.1080/14649365.2021.1876910.

Bhandar, Brenna. 2016. "Possession, Occupation and Registration: Recombinant Ownership in the Settler Colony." *Settler Colonial Studies* 6, no. 2 (April): 119–32. https://doi.org/10.1080/2201473X.2015.1024366.

Bickers, Robert A. 2010. *Settlers and Expatriates: Britons over the Seas*. Oxford: Oxford University Press.

Bigger, Patrick, Jessica Dempsey, Adeniyi P. Asiyanbi, Kelly Kay, Rebecca Lave, Becky Mansfield, Tracey Osborne, Morgan Robertson, and Gregory L. Simon. 2018. "Reflecting on Neoliberal Natures: An Exchange." *Environment and Planning E: Nature and Space* 1, nos. 1–2 (March–June): 25–75. https://doi.org/10.1177/2514848618776864.

Blair, James. 2017. "Settler Indigeneity and the Eradication of the Non-Native: Self-Determination and Biosecurity in the Falkland Islands (Malvinas)." *Journal of the Royal Anthropological Institute* 23, no. 3 (September): 580–602. https://doi.org/10.1111/1467-9655.12653.

Bluwstein, Jevgeniy. 2018. "From Colonial Fortresses to Neoliberal Landscapes in Northern Tanzania: A Biopolitical Ecology of Wildlife Conservation." *Journal of Political Ecology* 25 (1): 144–68. https://doi.org/10.2458/v25i1.22865.

– 2021a. "Colonizing Landscapes/Landscaping Colonies: From a Global History of Landscapism to the Contemporary Landscape Approach in Nature Conservation." *Journal of Political Ecology* 28 (1): 899–923. https://doi.org/10.2458/jpe.2850.

– 2021b. "Transformation is Not a Metaphor." *Political Geography* 90 (October): 102450. https://doi.org/10.1016/j.polgeo.2021.102450.

Boone, Catherine, Fibian Lukalo, and Sandra F. Joireman. 2021. "Promised land: settlement schemes in Kenya, 1962 to 2016." *Political Geography* 89 (August): 102393. https://doi.org/10.1016/j.polgeo.2021.102393.

Borana Conservancy (BC). n.d.-a. "Activities." Accessed 1 September 2021. https://www.borana.co.ke/activities.

– n.d.-b. *Borana Conservancy Newsletter III*. Nanyuki: Borana Conservancy.

Braverman, Irus. 2023. *Settling Nature: The Conservation Regime in Palestine-Israel*. Minneapolis: University of Minnesota Press.

Brockington, Dan. 2002. *Fortress conservation: The preservation of the Mkomazi Game Reserve, Tanzania*. Bloomington: Indiana University Press.

Brockington, Dan, Rosaleen Duffy, and Jim Igoe. 2012. *Nature Unbound: Conservation, Capitalism and the Future of Protected Areas*. London: Routledge.

Brockington, Daniel, and Jim Igoe. 2006. "Eviction for Conservation: A Global Overview." *Conservation and Society* 4, no. 3 (July–September): 424–70.

Browne, Peter. 2017. "Is This the Most Beautiful Safari Lodge in Africa?" Condé Nast Traveller, 4 January 2017. https://www.cntraveller.com/gallery/arijiju-kenya-safari-lodge.

Bull, Bartle. 1988. *Safari: A chronicle of Adventure*. London: Penguin.

Büscher, Bram. 2011. "The Neoliberalisation of Nature in Africa." In *African Engagements: Africa Negotiating an Emerging Multipolar World*, edited by Ton Dietz, Kjell Havnevik, Mayke Kaag, and Terje Oestigaard, 84–109. Leiden: Brill.

– 2021. "Between Overstocking and Extinction: Conservation and the Intensification of Uneven Wildlife Geographies in Africa." *Journal of Political Ecology* 28 (1): 1–22. https://doi.org/10.2458/jpe.2956.

Büscher, Bram, and Robert Fletcher. 2019. "Towards Convivial Conservation." *Conservation & Society* 17, no. 3 (July–September): 283–96. https://doi .org/10.4103/cs.cs_19_75.

Büscher, Bram, Robert Fletcher, Dan Brockington, Chris Sandbrook, Bill Adams, Lisa Campbell, Catherine Corson, et al. 2017. "Doing Whole Earth Justice: A Reply to Cafaro et al." *Oryx* 51, no. 3 (July): 401. https://doi .org/10.1017/s0030605317000278.

Büscher, Bram, Sian Sullivan, Katja Neves, Jim Igoe, and Dan Brockington. 2012. "Towards a Synthesized Critique of Neoliberal Biodiversity Conservation." *Capitalism Nature Socialism* 23, no. 2 (June): 4–30. https:// doi.org/10.1080/10455752.2012.674149.

Butt, Bilal. 2014. "The Political Ecology of 'Incursions': Livestock, Protected Areas and Socio-Ecological Dynamics in the Mara Region of Kenya." *Africa* 84, no. 4 (November): 614–37. https://doi.org/10.1017/s0001972014000515.

Butz, Ramona, Becky Estes, Michele Slaton, Carol Clark, and John Kerkering. 2018. *Opuntia Mapping and Monitoring in Kenya*. Washington, DC: USAID.

Byrd, Jodi A. 2011. *The Transit of Empire: Indigenous Critiques of Colonialism*. Minneapolis: University of Minnesota Press.

Cafaro, Philip, Tom Butler, Eileen Crist, Paul Cryer, Eric Dinerstein, Helen Kopnina, Reed Noss, et al. 2017. "If We Want a Whole Earth, Nature Needs Half: A Response to Büscher et al." *Oryx* 51, no. 3 (July): 400. https://doi .org/10.1017/s0030605317000072.

Capital FM. 2016. "Kenyan Conservationists Ian Craig Awarded Order of the British Empire." Capital Life Style, 15 June 2016. https://www.capitalfm .co.ke/lifestyle/2016/06/15/africa-kenya-wildlife-lewa-conservation -kenyan-conservationist-ian-craig-awarded-order-british-empire/.

Cariño, Joji, and Maurizio Farhan Ferrari. 2021. Negotiating the Futures of Nature and Cultures: Perspectives from Indigenous Peoples and Local Communities about the Post-2020 Global Biodiversity Framework." *Journal of Ethnobiology* 41, no. 2 (July): 192–208. https://doi.org/10.2993/0278-0771-41.2.192.

Carling, Jørgen, Marta Bivand Erdal, and Rojan Ezzati. 2014. "Beyond the Insider–Outsider Divide in Migration Research." *Migration Studies* 2, no. 1 (March): 36–54. https://doi.org/10.1093/migration/mnt022.

Cavanagh, Connor J., Teklehaymanot Weldemichel, and Tor A. Benjaminsen. 2020. "Gentrifying the African Landscape: The Performance and Powers of

For-Profit Conservation on Southern Kenya's Conservancy Frontier." *Annals of the American Association of Geographers* 110, no. 5 (September): 1594–612. https://doi.org/10.1080/24694452.2020.1723398.

Ceballos, Gerardo, Paul R. Ehrlich, Anthony D. Barnosky, Andrés García, Robert M. Pringle, and Todd M. Palmer. 2015. "Accelerated Modern Human–Induced Species Losses: Entering the Sixth Mass Extinction." *Science Advances* 1, no. 5 (June): e1400253. https://doi.org/10.1126/sciadv.1400253.

Ceballos, Gerardo, Paul R. Ehrlich, and Peter H. Raven. 2020. "Vertebrates on the Brink as Indicators of Biological Annihilation and the Sixth Mass Extinction." *Proceedings of the National Academy of Sciences* 117, no. 24 (1 June): 13596–602. https://doi.org/10.1073/pnas.1922686117.

Champion, Arthur M. 1933. "Soil Erosion in Africa." *The Geographical Journal* 82, no. 2 (August): 130–39. https://doi.org/10.2307/1785660.

Chongwa, Mungumi Bakari. 2012. "The History and Evolution of National Parks in Kenya." In *The George Wright Forum* 29 (1): 9–42.

Coe, Neil M., and David C. Jordhus-Lier. 2011. "Constrained Agency? Re-evaluating the Geographies of Labour." *Progress in Human Geography* 35, no. 2 (April): 211–33. https://doi.org/10.1177/0309132510366746.

Coghlan, David. 2007. "Insider Action Research Doctorates: Generating Actionable Knowledge." *Higher Education* 54, no. 2 (August): 293–306. https://doi.org/10.1007/s10734-005-5450-0.

Cohn, Jeffrey P. 1988. "Halting the Rhino's Demise." *Bioscience* 38, no. 11 (December): 740–44. https://doi.org/10.2307/1310780.

Collard, Rosemary C. 2020. *Animal Traffic: Lively Capital in the Global Exotic Pet Trade*. Durham: Duke University Press.

Collett, David. 1987. Pastoralists and Wildlife: Image and Reality in Kenya Maasailand. In *Conservation in Africa: People, Policies, and Practice*, edited by David Anderson and Richard H. Grove, 129–48. Cambridge: Cambridge University Press.

Copley, Hugh. 1940. "Trout in Kenya Colony: Part I – Brown Trout." *The East African Agricultural Journal* 5, no. 5 (March): 345–61. https://doi.org/10.1080/03670074.1940.11663992.

Cottar Safari Service. 1937. *Big Game Hunting in Africa & Asia*. Nairobi: Cottar Safari Service.

Crosby, Alfred W. 1989. *Ecological Imperialism: The Biological Expansion of Europe, 900–1900*. Cambridge: Cambridge University Press.

Dahir, Abdi Latif. 2017. "Kenya's Opposition Leader Wants to Dismantle White-Owned Ranches." *Quartz Africa*, 13 June 2017. https://qz.com/africa/1004548/kenyas-raila-odinga-says-he-will-dismantle-white-owned-ranches-in-laikipia-if-he-wins-the-2017-election/.

Dar, Javid Ahmad, Subashree Kothandaraman, Pramod Kumar Khare, and Mohammed Latif Khan. 2022. "Sacred Groves of Central India: Diversity

Status, Carbon Storage, and Conservation Strategies." *Biotropica* 54, no. 6 (November): 1400–11. https://doi.org/10.1111/btp.13157.

D'Arcangelis, Carol L. 2018. "Revelations of a White Settler Woman Scholar-Activist: The Fraught Promise of Self-Reflexivity." *Cultural Studies ↔ Critical Methodologies* 18, no. 5 (October): 339–53. https://doi.org/10.1177/1532708617750675.

Day, Iyko. 2015. "Being or Nothingness: Indigeneity, Antiblackness, and Settler Colonial Critique." *Critical Ethnic Studies* 1, no. 2 (Fall): 102–21. https://doi.org/10.5749/jcritethnstud.1.2.0102.

Denzin, Norman K. 2019. "Grounded Theory and the Politics of Interpretation, Redux." *The SAGE Handbook of Current Developments in Grounded Theory*, edited by Antony Bryant and Kathy Charmaz, 454–71. Thousand Oaks: Sage Publications. https://doi.org/10.4135/9781526485656.

Department of Agriculture (DOA). 1930. *The Evils of Soil Erosion and Ways of Preserving the Land*. Nairobi: The Soil Conservation Services, Department of Agriculture, Kenya Colony.

Díaz, Sandra Myrna, Josef Settele, Eduardo Brondízio, Hien T. Ngo, Maximilien Guèze, John Agard, Almut Arneth, et al. 2019. "The Global Assessment Report on Biodiversity and Ecosystem Services: Summary for Policy Makers." In *Summary for Policymakers of the Global Assessment Report on Biodiversity and Ecosystem Eervices of the Intergovernmental Science-Policy Platform on Biodiversity and Ecosystem Services*. Bonn: IPBES Secretariat.

Dicenta, Mara. 2023. "White Animals: Racializing Sheep and Beavers in the Argentinian Tierra del Fuego." *Latin American and Caribbean Ethnic Studies* 18, no. 2 (April): 308–29. https://doi.org/10.1080/17442222.2021.2015140.

Didier, Karl A., Alayne Cotterill, Iain Douglas-Hamilton, Laurence Frank, Nicholas J. Georgiadis, Max Graham, Festus Ihwagi, Juliet King, Delphine Malleret-King, Dan Rubenstein, et al. 2011. "Landscape-Scale Conservation Planning of the Ewaso Nyiro: A Model for Land Use Planning in Kenya?" *Smithsonian Contributions to Zoology*, no. 632: 105–32.

Dlamini, Jacob S.T. 2020. *Safari Nation: A Social History of the Kruger National Park*. Athens: Ohio University Press.

Donovan, Josephine. 2006. "Feminism and the Treatment of Animals: From Care to Dialogue." *Signs: Journal of Women in Culture and Society* 31, no. 2 (Winter): 305–29. https://doi.org/10.1086/491750.

Doro, Marion E. 1979. "'Human Souvenirs of Another Era': Europeans in Post-Kenyatta Kenya." *Africa Today* 26 (3): 43–54.

Dressler, Wolfram, Bram Büscher, Michael Schoon, D.A.N. Brockington, Tanya Hayes, Christian A. Kull, James McCarthy, and Krishna Shrestha. 2010. "From Hope to Crisis and Back Again? A Critical History of the Global CBNRM Narrative." *Environmental Conservation* 37, no. 1 (March): 5–15. https://doi.org/10.1017/s0376892910000044.

Duder, C.J.D. 1989. "The Settler Response to the Indian Crisis of 1923 in Kenya: Brigadier General Philip Wheatley and 'Direct Action.'" *The Journal of Imperial and Commonwealth History* 17, no. 3 (May): 349–73. https://doi .org/10.1080/03086538908582797.

Duder, C.J.D., and Christopher P. Youé. 1994. "Paice's Place: Race and Politics in Nanyuki District, Kenya, in the 1920s." *African Affairs* 93, no. 371 (April): 253–78. https://doi.org/10.1093/oxfordjournals.afraf.a098711.

Duder, C.J.D., and G.L. Simpson. 1997. "Land and Murder in Colonial Kenya: The Leroghi Land Dispute and the Powys 'Murder' Case." *The Journal of Imperial and Commonwealth History* 25, no. 3 (September): 440–65. https:// doi.org/10.1080/03086539708583008.

Duffy, Rosaleen. 2014. "Waging a War to Save Biodiversity: The Rise of Militarized Conservation." *International Affairs* 90, no. 4 (July): 819–34. https://doi.org/10.1111/1468-2346.12142.

Duffy, Rosaleen, Francis Massé, Emile Smidt, Esther Marijnen, Bram Büscher, Judith Verweijen, Maano Ramutsindela, Trishant Simlai, Laure Joanny, and Elizabeth Lunstrum. 2019. "Why We Must Question the Militarisation of Conservation." *Biological Conservation* 232 (April): 66–73. https://doi .org/10.1016/j.biocon.2019.01.013.

Dwyer, Sonya Corbin, and Jennifer L. Buckle. 2009. "The Space Between: On Being an Insider-Outsider in Qualitative Research." *International Journal of Qualitative Methods* 8, no. 1 (March): 54–63. https://doi.org/10.1177 /160940690900800105.

East Africa Women's League (EAWL). 1953. *Are You Coming to Kenya? A Guide for the Woman Settler*. Nairobi: EAWL.

Ebru News. 2017. "Bullets Used by Pokots in Laikipia Raids Are from State's Eldoret Artillery." Ebru Television, 23 April 2017. https://ebru.co.ke /bullets-used-by-pokots-in-laikipia-raids-are-from-states-eldoret-artillery/.

Eliot, Charles. 1905. *The East Africa Protectorate*. London: Arnold.

Elliott, Joanna, and Muthoni M. Mwangi. 1997. "Making Wildlife 'Pay' in Laikipia, Kenya. Laikipia Wildlife Economics Study Discussion Paper CEC-DP-1." Nairobi: African Wildlife Foundation.

Expert Africa. n.d. "Golden Jackal Fly-In Safari." Accessed 1 September 2021. https://www.expertafrica.com/kenya/safari/golden-jackal-fly-in-safari /in-detail.

Fanon, Frantz. 1963. *The Wretched of the Earth*. New York: Grove Press.

Fanstone, Ben Paul. 2016. "The Pursuit of the 'Good Forest' in Kenya, c. 1890– 1963: The History of the Contested Development of State Forestry within a Colonial Settler State." PhD diss., University of Stirling, Stirling, UK.

Fitch, Chris. 2016. "Lewa: A Model for Kenyan Conservation." *Geographical*, 22 August 2016. Archived 29 August 2016, at the Wayback Machine, https:// web.archive.org/web/20160823164012/http://geographical.co.uk/nature /wildlife/item/1854-lewa-a-model-of-conservation.

Fitzsimons, James, Ian Pulsford, and Geoff Wescott. 2013. "Lessons from Large-Scale Conservation Networks in Australia." *Parks* 19, no. 1 (March): 115–25. https://doi.org/10.2305/iucn.ch.2013.parks-19-1.jf.en.

Fitzsimons, James A., and Geoff Wescott. 2005. "History and Attributes of Selected Australian Multi-Tenure Reserve Networks." *Australian Geographer* 36, no. 1 (March): 75–93. https://doi.org/10.1080/00049180500050904.

Fletcher, Robert. 2023. *Failing Forward: The Rise and Fall of Neoliberal Conservation*. Berkeley: University of California Press.

Food and Agriculture Organization of the UN (FAO), and United Nations Environment Programme (UNEP). 2020. *The State of the World's Forests: Forests, Biodiversity and People*. Rome: FAO.

Fox, Graham R. 2018a. "The 2017 Shooting of Kuki Gallmann and the Politics of Conservation in Northern Kenya." *African Studies Review* 61, no. 2 (July): 210–36. https://doi.org/10.1017/asr.2017.130.

– 2018b. "Maasai Group Ranches, Minority Land Owners, and the Political Landscape of Laikipia County, Kenya." *Journal of Eastern African Studies* 12, no. 3 (July): 473–93. https://doi.org/10.1080/17531055.2018.1471289.

Fryxell, J.M., and A.R.E. Sinclair. 1988. "Causes and Consequences of Migration by Large Herbivores." *Trends in Ecology & Evolution* 3, no. 9 (September): 237–41. https://doi.org/10.1016/0169-5347(88)90166-8.

Fung, Amy. 2021. "Is Settler Colonialism Just Another Study of Whiteness?" *Canadian Ethnic Studies* 53 (2): 115–31. https://doi.org/10.1353/ces.2021.0011.

Gaarder, Emily. 2011. *Women and the Animal Rights Movement*. New Brunswick: Rutgers University Press.

Gallman, Kuki. 1991. *I Dreamed of Africa*. London: Penguin.

Garland, Elizabeth. 2008. "The Elephant in the Room: Confronting the Colonial Character of Wildlife Conservation in Africa." *African Studies Review* 51, no. 3 (December): 51–74. https://doi.org/10.1353/arw.0.0095.

Garlick, Ben, and Kate Symons. 2020. "Geographies of Extinction: Exploring the Spatiotemporal Relations of Species Death." *Environmental Humanities* 12, no. 1 (May): 296–320. https://doi.org/10.1215/22011919-8142374.

Georgiadis, Nicholas. 1997. "Numbers and Distribution of Large Herbivores in Laikipia District: Sample Counts for February and September 1996." Nanyuki: Laikipia Wildlife Forum.

–, ed. 2011. *Conserving Wildlife in African Landscapes: Kenya's Ewaso Ecosystem*. Washington: Smithsonian Institution Scholarly Press.

Gibbs, James. 2014. "*Uhuru na Kenyatta*: White Settlers and the Symbolism of Kenya's Independence Day Events." *The Journal of Imperial and Commonwealth History* 42, no. 3 (May): 503–29. https://doi.org/10.1080/03086534.2014.901007.

Giesen, Wim, Paul Giesen, and Kris Giesen. 2006. "Habitat Changes at Lewa Wildlife Conservancy, Kenya: From Cattle Ranch to Conservation Area: Effects of Changing Management on Habitat from 1962–2006." Unpublished manuscript.

– 2017. "Lewa Wildlife Conservancy Habitat Changes 1962–2016." Unpublished manuscript.

Gillespie, Kathryn, and Yamini Narayanan. 2020. "Animal Nationalisms: Multispecies Cultural Politics, Race, and the (Un)Making of the Settler Nation-State." *Journal of Intercultural Studies* 41, no. 1 (January): 1–7. https://doi.org/10.1080/07256868.2019.1704379.

Gilliland, Haley. 2019. "Bold Effort to Save Rhino Completes Critical Step." *National Geographic*, 23 August 2019. https://www.nationalgeographic .com/animals/2019/08/plan-save-northern-white-rhino-ivf/.

Gissibl, Bernhard. 2016. *The Nature of German Imperialism: Conservation and the Politics of Wildlife in Colonial East Africa*. Oxford: Berghahn Books.

Gjefsen, Torbjørn. 2021. "Indigenous People Get Less than 1% of Climate Funding? It's Actually Worse (Commentary)." *Mongabay*, 19 November 2021. https://news.mongabay.com/2021/11/indigenous-people-get-less -than-1-of-climate-funding-its-actually-worse-commentary/.

Goldman, Mara. 2009. "Constructing Connectivity: Conservation Corridors and Conservation Politics in East African Rangelands." *Annals of the Association of American Geographers* 99, no. 2 (April): 335–59. https://doi .org/10.1080/00045600802708325.

Goudge, Paulette. 2003. *The Whiteness of Power: Racism in Third World Development and Aid*. London: Lawrence & Wishart.

Government of Kenya (GoK). 2017. *Wildlife Migratory Corridors and Dispersal Areas: Kenya Rangelands and Coastal Terrestrial Ecosystems. Kenya Vision 2030 Flagship Project: Securing Wildlife Migratory Routes and Corridors*. Nairobi: Ministry of Environment and Natural Resources, Directorate of Resource Surveys and Remote Sensing and Kenya Wildlife Service.

Graham, Max. 2001. "At Large in a Sea of Troubles." *Swara* 24, no. 2 (May–August): 38–40.

Green, Sian. 2016. "Movement and Distribution of the African Elephant Loxodonta Africana within the Mount Kenya Elephant Corridor." Master's thesis, University of Southampton, Southampton, UK.

Griffiths, Tom, and Libby Robin, eds. 1997. *Ecology and Empire: Environmental History of Settler Societies*. Washington: University of Washington Press.

Griffiths, Mark, Fridah Mueni, Kate Baker, and Surshti Patel. 2023. "Decolonising Spaces of Knowledge Production: Mpala Research Centre in Laikipia County, Kenya." *Environment and Planning E: Nature and Space* 6, no. 4 (December): 2340–57. https://doi.org/10.1177/25148486231156728.

Haggard, John. 2017. "Kenyan MP Declares 'Third World War' to Drive Out White Farmers." *The Sunday Times*, 12 March 2017. https://www.thetimes.co.uk/article/kenyan-mp-declares-third-world-war-to-drive-out-white-farmers-cscl3q23k.

Hall, Stuart. 1990. "Cultural Identity and Diaspora." In *Colonial Discourse and Post-Colonial Theory: A Reader*, edited by Patrick R.J. Williams and Laura Chrisman, 227–37. London: Harvester Wheatsheaf.

Hammer, Joshua. 1994. "Richard Leakey's Fall from Grace." *Outside Magazine*, June 1994. https ://www.outsideonline.com/outdoor-adventure/richard-leakeys-fall-grace/.

Haraway, Donna. 1990. "A Manifesto for Cyborgs: Science, Technology, and Socialist Feminism in the 1980s." In *Feminism/Postmodernism*, edited by Linda Nicholson, 190–233. London: Routledge.

Hardin, Garrett. 1968. "The Tragedy of the Commons: The Population Problem Has No Technical Solution; It Requires a Fundamental Extension in Morality." *Science* 162, no. 3859 (13 December): 1243–48. https://doi.org/10.1126/science.162.3859.1243.

Heath, Brian. 2000. "Ranching: An Economic Yardstick." In *Wildlife Conservation by Sustainable Use*, edited by Prins, Herbert H.T., Jan Geu Grootenhuis, and Thomas T. Dolan, 21–33. Boston: Kluwer Academic Publishers.

Helping Rhinos. n.d. "Rhino Strongholds." Accessed 17 May 2023. https://www.helpingrhinos.org/rhinostrongholds/.

Herne, Brian. 1999. *White Hunters: The Golden Age of African Safaris*. New York: Henry Holt and Company.

Heydinger, John. 2021. "Human-Lion Conflict and the Reproduction of White Supremacy in Northwest Namibia." *African Studies Review* 64, no. 4 (December): 909–37. https://doi.org/10.1017/asr.2021.72.

Hilty, Jodi, Graeme L. Worboys, Annika Keeley, Stephen Woodley, Barbara Lausche, Harvey Locke, Mark Carr, et al. 2020. *Guidelines for Conserving Connectivity through Ecological Networks and Corridors*. Best Practice Protected Area Guidelines Series 30, edited by Craig Groves. Gland: IUCN. https://doi.org/10.2305/IUCN.CH.2020.PAG.30.en.

Hingston, Richard W. G. 1931. "Proposed British National Parks for Africa." *The Geographical Journal* 77, no. 5 (May): 401–22. https://doi.org/10.2307/1783602.

Hobley, Charles W. 1933. "Soil Rrosion: A Problem in Human Geography." *The Geographical Journal* 82, no. 2 (August): 139–46. https://doi.org/10.2307/1785661.

Hughes, Lotte. 2006. *Moving the Maasai: A Colonial Misadventure*. New York: Springer.

Hugill, David. 2017. "What Is a Settler-Colonial City?" *Geography Compass* 11, no. 5 (May): e12315. https://doi.org/10.1111/gec3.12315.

Huxley, Elspeth. 1937. "The Menace of Soil Rrosion." *Journal of the Royal African Society* 36, no. 144 (July): 357–70. https://doi.org/10.1093/oxfordjournals.afraf.a101460.

– 1948. *Settlers of Kenya*. Westport: Greenwood.

– 1953. *White Man's Country: Lord Delamere and the Making of Kenya. Vol I, 1870–1914*. London: Chatto and Windus.

– 1959. *The Flame Trees of Thika*. London: Vintage Books.

– 1991. *Nine Faces of Kenya*. New York: Viking.

Igoe, James. 2017. *The Nature of Spectacle: On Images, Money, and Conserving Capitalism*. Tucson: University of Arizona Press.

Infonet Biovision. n.d. "Brown Olive." Infonet. Accessed 16 October 2023. https://www.infonet-biovision.org/EnvironmentalHealth/Trees/brown-olive.

International Union for the Conservation of Nature (IUCN), and Species Survival Commission (SSC). 2012. *IUCN Guidelines for Reintroductions and Other Conservation Translocations*. Gland: IUCN-SCC.

Izuakor, Levi I. 1988. "The Environment of Unreality: Nurturing a European Settlement in Kenya." *Journal of Asian and African Studies* 23 (3–4): 317–24.

Jackson, Will. 2011. "White Man's Country: Kenya Colony and the Making of a Myth." *Journal of East African Studies* 5, no. 2 (May): 344–68. https://doi.org/10.1080/17531055.2011.571393.

Jepson, Paul, and Robert J. Whittaker. 2002. "Histories of Protected Areas: Internationalisation of Conservationist Values and Their Adoption in the Netherlands Indies (Indonesia)." *Environment and History* 8, no. 2 (May): 129–72. https://doi.org/10.3197/096734002129342620.

Johns, David. 2019. "History of Rewilding: Ideas and Practice." In *Rewilding*, edited by Nathalie Pettorelli, Sarah M. Durant, and Johan T. Du Toit, 12–33. Cambridge: Cambridge University Press.

Johnston, Anna, and Alan Lawson. 2000. "Settler Colonies." In *A Companion to Postcolonial Studies*, edited by Henry Schwarz and Sangeeta Ray, 360–76. Boston: Blackwell.

Jones, Will. 2019. "Sera and Melako: The NRT and the Community Conservancy as Excellent Conservation Model." *Wild Philanthropy*, 14 June 2019. https://wildphilanthropy.com/sera-and-melako-the-nrt-and-the-community-conservancy-as-excellent-conservation-model/.

Jørgensen, Dolly. 2015. "Rethinking Rewilding." *Geoforum* 65 (October): 482–8. https://doi.org/10.1016/j.geoforum.2014.11.016.

Joudah, Nour. 2020. "Intervention – 'Gaza as Site and Method: The Settler Colonial City Without Settlers.'" *Antipode*, 24 August 2020. https://antipodeonline.org/2020/08/24/gaza-as-site-and-method/.

Kabiri, Ngeta. 2010. "Historic and Contemporary Struggles for a Local Wildlife Governance Regime in Kenya." In *Community Rights, Conservation, and Contested Land: The Politics of Natural Resource Governance in Africa*, edited by Fred Nelson, 121–46. London: Earthscan.

Kahurananga, James, and Frank Silkiluwasha. 1997. "The Migration of Zebra and Wildebeest between Tarangire National Park and Simanjiro Plains, Northern Tanzania, in 1972 and Recent Trends." *African Journal of Ecology* 35, no. 3 (September): 179–85. https://doi.org/10.1111/j.1365-2028.1997.071-89071.x.

Kamau, J., Wycliffe Kiiya, Sammy Ajanga, Nasirembe Wanyonyi, Geoffrey Gathungu, Mabel Mahasi, and E. Pertet. 2019. *Pyrethrum Propagation*. Edited by Lusike Wasilwa. Nairobi: Kenya Agricultural & Livestock Research Organization.

Kamuti, Tariro. 2014. "The Fractured State in the Governance of Private Game Farming: The Case of KwaZulu-Natal Province, South Africa." *Journal of Contemporary African Studies* 32, no. 2 (April): 190–206. https://doi.org/10.1080/02589001.2014.936678.

Kantai, Parselelo. 2007. "In the Grip of the Vampire State: Maasai Land Struggles in Kenyan Politics." *Journal of Eastern African Studies* 1, no. 1 (March): 107–22. https://doi.org/10.1080/17531050701218890.

Kanyinga, Karuti. 2009. "The Legacy of the White Highlands: Land Rights, Ethnicity and the Post-2007 Election Violence in Kenya." *Journal of Contemporary African Studies* 27, no. 3 (July): 325–44. https://doi.org/10.1080/02589000903154834.

Karuri, Alice Nyawira. 2021. "Adaptation of Small-Scale Tea and Coffee Farmers in Kenya to Climate Change." In *African Handbook of Climate Change Adaptation*, edited by Walter Leal Filho, Nicholas Oguge, Desalegn Ayal, Lydia Adeleke, and Izael da Silva, 22–47. New York: Springer.

Kauanui, Kēhaulani J. 2017. "'A Structure, Not an Event': Settler Colonialism and Enduring Indigeneity." *Lateral* 5, no. 1 (Spring). https://doi.org/10.25158/l5.1.7.

Kenya Human Rights Commission (KHRC). 2012. "MAU MAU Case: Dealing with Past Colonial Injustices." Press release, 23 July 2012. https://www.khrc.or.ke/2015-03-04-10-37-01/press-releases/454-mau-mau-case-dealing-with-past-colonial-injustices.html.

Kenya Land Commission. 1933. *Kenya Land Commission Evidence*, vol. 1. Nairobi: Government Printer.

Kenya National Bureau of Statistics (KNBS). 2019. *Kenya Population and Housing Census Volume IV: Distribution of Population by Socio-Economic Characteristics*. Nairobi: Kenya National Bureau of Statistics.

Kenya Wildlife Conservancies Association (KWCA). n.d. "Facts and Figures." Conservancies. Accessed 19 October 2021. https://kwcakenya.com/conservancies/status-of-wildlife-conservancies-in-kenya/.

Kenya Wildlife Service (KWS). 2019a. "KWS Launches the National Recovery and Action Plan for the Mountain Bongo (2019–2023)." 10 July 2019.

https://www.kws.go.ke/content/kws-launches-national-recovery-and
-action-plan-mountain-bongo-2019-2023.

– 2019b. *National Recovery and Action Plan for the Mountain Bongo (Tragelaphus eurycerus isaaci) in Kenya (2019–2023)*. Nariobi: Republic of Kenya.

Kepe, Thembela. 2009. "Shaped by Race: Why 'Race' Still Matters in the Challenges Facing Biodiversity Conservation in Africa." *Local Environment* 14, no. 9 (October): 871–8. https://doi.org/10.1080/13549830903164185.

Khayale, Cedric, Linus Kariuki, Geoffrey Chege, Mxolisi Sibanda, Martin Mulama, Benson Okita-Ouma, and Rajan Amin. 2020. "Progress on the Kenya Black Rhino Action Plan (2017–2021)." *Pachyderm* 61 (June): 109–19.

Kimiti, David, Mary Mwololo, Timothi Kaaria, Ian Lemaiyan, Saibala Gilisho, Francis Kobia, and Edwin Kisio. 2017. *Research and Monitoring Annual Report 2017*. Isiolo: Research Department of the Lewa-Borana Landscape.

Kimiti, David W., Geoffrey Chege, Ian Lemaiyan, and Lara Jackson. 2017. *When Less is More: Adapting Black Rhino Conservation Targets in Response to Long-Term Ecological and Population Data*. Isiolo: Lewa Wildlife Conservancy.

Kimiti, David W., Timothy Kaaria, Edwin Kisio, Kenneth Onzere, Eunice Kamau, Saibala Gilicho, Francis Kobia, and Geoffrey Chege. 2019. *Research and Monitoring: Annual Report 2019*. Isiolo: Lewa Wildlife Conservancy.

Kinnaird, Margaret F., and Timothy G. O'Brien. 2012. "Effects of Private-Land Use, Livestock Management, and Human Tolerance on Diversity, Distribution, and Abundance of Large African Mammals." *Conservation Biology* 26, no. 6 (December): 1026–39. https://doi.org/10.1111/j.1523-1739.2012.01942.x.

Kirathe, Joseph Nderitu, John Maina Githaiga, Robert Mutugi Chira, and Daniel Rubenstein. 2021. "Land Use Influence on Distribution and Abundance of Herbivores in Samburu-Laikipia, Kenya." *Journal of Sustainability, Environment and Peace* 4, no. 1 (July): 21–9. https://doi.org/10.53537/jsep.2021.07.003.

Knight, Tim. 2019. "Special Delivery – Baby Rhino Bonus for Community-Run Conservancy in Kenya. Fauna and Flora International, 14 October 2019. https://www.fauna-flora.org/news/special-delivery-baby-rhino-bonus-community-run-conservancy-kenya.

Knight, Wanda B., and Yang Deng. 2016. "N/either Here N/or There: Culture, Location, Positionality, and Art Education." *Visual Arts Research* 42, no. 2 (Winter): 105–11. https://doi.org/10.5406/visuartsrese.42.2.0105.

Kock, Richard A. 1995. "Wildlife Utilization: Use It or Lose It – A Kenyan Perspective." *Biodiversity & Conservation* 4, no. 3 (April): 241–56. https://doi.org/10.1007/bf00055971.

Koot, Stasja, Bram Büscher, and Lerato Thakholi. 2022. "The New Green Apartheid? Race, Capital, and Logics of Enclosure in South Africa's Wildlife Economy." *Environment and Planning E: Nature and Space*. Published OnlineFirst, 28 June 2022. https://doi.org/10.1177/25148486221110438.

Kruse, Corwin R. 1999. "Gender, Views of Nature, and Support for Animal Rights." *Society & Animals* 7 (3): 179–98. https://doi.org/10.1163/156853099x00077.

Laffrey, Anna. 2018. "'Mama Elephant': How Daphne Sheldrick Changed the Fate of Elephants Worldwide." *CNN*, 5 August 2018. https://edition.cnn.com/2018/08/15/africa/kenya-daphne-sheldrick-wildlife-conservation/index.html.

Laikipia Forum. n.d. "Land Use & Management." Accessed 19 October 2021. https://laikipia.org/land-use-management/.

Laikipia Wildlife Forum (LWF). 2012. *Wildlife Conservation Strategy for Laikipia County: 2012–2030*. Nanyuki: Laikipia Wildlife Forum.

Lalampaa, Tom. 2021. "Response to an Editorial Column Published in the Daily Nation – "Stop Conservancies in North to Protect Community Lands There" by Kaltum Guyo." Northern Rangelands Trust, 5 July 2021. https://www.nrt-kenya.org/news-2/2021/7/14/response-to-an-article-published-in-the-daily-nation-stop-conservancies-in-north-to-protect-community-lands-there-by-kaltum-gayo.

Laltaika, Elifuraha I., and Kelly M. Askew. 2021. "Modes of Dispossession of Indigenous Lands and Territories in Africa." In *Lands of the Future: Anthropological Perspectives on Pastoralism, Land Deals and Tropes of Modernity in Eastern Africa*, edited by Echi Christina Gabbert, Fana Gebresenbet, John G. Galaty, and Günther Schlee, 99–122. New York: Berghahn Books.

Lawrence, Bonita, and Enakshi Dua. 2005. "Decolonizing Antiracism." *Social Justice* 32, no. 4 (102): 120–43.

Lenzner, Bernd, Guillaume Latombe, Anna Schertler, Hanno Seebens, Qiang Yang, Marten Winter, Patrick Weigelt, et al. 2022. "Naturalized Alien Floras Still Carry the Legacy of European Colonialism." *Nature Ecology & Evolution* 6, no. 11 (November): 1723–32. https://doi.org/10.1038/s41559-022-01865-1.

Letai, John. 2015. "Land Deals and Pastoralist Livelihoods in Laikipia County, Kenya." In *Africa's Land Rush: Rural Livelihoods and Agrarian Change*, edited by Ruth Hall, Ian Scoones, and Dzodzi Tsikata, 83–98. Suffolk: James Currey.

Levins, Richard. 1969. "Some Demographic and Genetic Consequences of Environmental Heterogeneity for Biological Control." *American Entomologist* 15, no. 3 (September): 237–40. https://doi.org/10.1093/besa/15.3.237.

Lewa Wildlife Conservancy (LWC). n.d. "Adopting a Rhino." Accessed 23 July 2021. https://www.lewa.org/adopt-a-rhino/.

– 2017. *Impact Report 2017*. Isiolo: Lewa Wildlife Conservancy.

Li, Tania M. 2014. "Fixing Non-market Subjects: Governing Land and Population in the Global South." *Foucault Studies* 18 (October): 34–48. https://doi.org/10.22439/fs.v0i18.4650.

Lipscomb, Joseph. 1955. *White Africans*. London: Faber and Faber.

Litoroh, Moses, Patrick Omondi, Richard Kock, and Rajan Amin. 2012. *Conservation and Management Strategy for the Elephant in Kenya: 2012–2021*. Nairobi: KWS.

Loarie, Scott R., Rudi J. Van Aarde, and Stuart L. Pimm. 2009. "Fences and Artificial Water Affect African Savannah Elephant Movement Patterns." *Biological conservation* 142, no. 12 (December): 3086–98. https://doi .org/10.1016/j.biocon.2009.08.008.

Lolwerikoi, Michael Lmatila. 2010. *Orality and the Land: The Impact of Colonialism on Lmaa Narratives in Kenya*. Diss., Asbury Theological Seminary.

Loring, Philip A., and Faisal Moola. 2021. "Erasure of Indigenous Peoples Risks Perpetuating Conservation's Colonial Harms and Undermining its Future Effectiveness." *Conservation Letters* 14, no. 2 (March–April): e12782. https://doi.org/10.1111/conl.12782.

Løvschal, Mette, and Marie Ladekjær Gravesen. 2021. "De-/Fencing Grasslands: Ongoing Boundary Making and Unmaking in Postcolonial Kenya." *Land* 10, no. 8 (August): 786. https://doi.org/10.3390/land10080786.

Lunstrum, Elizabeth. 2018. "Capitalism, Wealth, and Conservation in the Age of Security: The Vitalization of the State." *Annals of the American Association of Geographers* 108, no. 4 (July): 1022–37. https://doi.org/10.1080/24694452 .2017.1407629.

Lunstrum, Elizabeth, and Megan Ybarra. 2018. "Deploying Difference: Security Threat Narratives and State Displacement from Protected Areas." *Conservation and Society* 16, no. 2 (April–June): 114–24. https://doi.org /10.4103/cs.cs_16_119.

Mabele, Mathew Bukhi, Laila Thomaz Sandroni, Y. Ariadne Collins, and June Rubis. 2021. "What Do We Mean by Decolonizing Conservation? A Response to Lanjouw 2021." CONVIVA, 27 October 2021. Archived 20 October 2023, at the Wayback Machine, https://web.archive.org /web/20231020224455/https://conviva-research.com/what-do-we-mean -by-decolonizing-conservation-a-response-to-lanjouw-2021/.

MacArthur, Robert H., and Edward O. Wilson. 1967. *The Theory of Island Biogeography*. Princeton: Princeton University Press.

MacEacheran, Mike. 2013. "The Rhino Wars of Central Kenya." *BBC*, 15 August 2013. https://www.bbc.com/travel/article/20130812-the-rhino -wars-of-central-kenya.

MacKenzie, John M. 1998. "Empire and National Identities: The Case of Scotland." *Transactions of the Royal Historical Society* 8 (December): 215–31. https://doi.org/10.2307/3679295.

Makanda, D.W., and J.F. Oehmke. 1995. "Kenya's Wheat Agriculture: Past, Present and Future." Staff paper #95–54. Department of Agricultural Economics, Michigan State University.

Maldonado-Torres, Nelson. 2007. "On the Coloniality of Being: Contributions to the Development of a Concept." *Cultural studies* 21, nos. 2–3 (March): 240–70. https://doi.org/10.1080/09502380601162548.

Mamdani, Mahmood. 1996. "Indirect Rule, Civil Society, and Ethnicity: The African Dilemma." *Social Justice* 23, nos. 1–2 (63–4, Spring–Summer): 145–50.

– 2006. *Citizen and Subject: Contemporary Africa and the Legacy of Late Colonialism*. Princeton: Princeton University Press.

Mar, Tracey Banivanua, and Penelope Edmonds. 2010. "Introduction: Making Space in Settler Colonies." In *Making Settler Colonial Space*, edited by Tracey Banivanua Mar and Penelope Edmonds, 1–24. London: Palgrave Macmillan.

Maréchal, Laëtitia, Ann MacLarnon, Bonaventura Majolo, and Stuart Semple. 2016. "Primates' Behavioural Responses to Tourists: Evidence for a Trade-Off between Potential Risks and Benefits." *Scientific Reports* 6 (1): 32465.

Margulies, Jared D., and Brock Bersaglio. 2018. "Furthering Post-human Political Ecologies. *Geoforum* 94 (August): 103–6. https://doi.org/10.1016/j.geoforum.2018.03.017.

Marshall, Sarah. 2021. "Segera Retreat: Luxury Safari Lodge in Northern Kenya." *The Sunday Times*, 24 April 2021. https://www.thetimes.co.uk/article/segera-retreat-luxury-safari-lodge-in-northern-kenya-swxfnl2rx.

Matheka, Ruben. 2005. "Antecedents to the Community Wildlife Conservation Programme in Kenya, 1946–1964." *Environment and History* 11, no. 3 (August): 239–67. https://doi.org/10.3197/096734005774434539.

– 2008. "Decolonisation and Wildlife Conservation in Kenya, 1958–68." *The Journal of Imperial and Commonwealth History* 36, no. 4 (December): 615–39. https://doi.org/10.1080/03086530802561016.

Mbaria, Gitau. 2017. "The Laikipia Crisis and the Disenfranchisement of Kenyans in the North." The Elephant, 18 May 2017. https://www.theelephant.info/features/2017/05/18/the-laikipia-crisis-and-the-disenfranchisement-of-kenyans-in-the-north/.

Mbaria, John, and Mordecai Ogada. 2016. *The Big Conservation Lie: The Untold Story of Wildlife Conservation in Kenya*. Auburn: Lens&Pens.

McClanahan, Young, Truman P. Young, and Tim McClanahan. 1996. *East African Ecosystems and Their Conservation*. Oxford: Oxford University Press.

McGregor, JoAnn. 2005. "Crocodile Crimes: People versus Wildlife and the Politics of Postcolonial Conservation on Lake Kariba, Zimbabwe." *Geoforum* 36, no. 3 (May): 353–69. https://doi.org/10.1016/j.geoforum.2004.06.007.

McIntosh, Janet. 2016. *Unsettled: Denial and Belonging among White Kenyans*. Oakland: University of California Press.

– 2017. "Land, Belonging and Structural Oblivion among Contemporary White Kenyans." *Africa* 87, no. 4 (November): 662–82. https://doi.org/10.1017/s0001972017000304.

McLeod, John. 2010. *Case Study Research in Counselling and Psychotherapy.* Thousand Oaks: Sage Publications.

Melubo, Kokel. 2020. "Why Are Wildlife on the Maasai Doorsteps? Insights from the Maasai of Tanzania." *AlterNative: An International Journal of Indigenous Peoples* 16, no. 3 (September): 180–92. https://doi .org/10.1177/1177180120947823.

Miller, Robert J., Jacinta Ruru, Larissa Behrendt, and Tracey Lindberg. 2010. *Discovering Indigenous Lands: The Doctrine of Discovery in the English Colonies.* Oxford: Oxford University Press.

Mizutani, Fumi. 1999. "Biomass Density of Wild and Domestic Herbivores and Carrying Capacity on a Working Ranch in Laikipia District, Kenya." *African Journal of Ecology* 37, no. 2 (June): 226–40. https://doi.org/10.1046 /j.1365-2028.1999.00171.x.

Mollett, Sharlene. 2017. "Irreconcilable Differences? A Postcolonial Intersectional Reading of Gender, Development and Human Rights in Latin America." *Gender, Place & Culture* 24, no. 1 (January): 1–17. https://doi.org /10.1080/0966369x.2017.1277292.

Morgan, W. T. W. 1963. "The 'White Highlands' of Kenya." *The Geographical Journal* 129, no. 2 (June): 140–55. https://doi.org/10.2307/1792632.

Nacoulma, Blandine Marie Ivette, Katharina Schumann, Salifou Traoré, Markus Bernhardt-Römermann, Karen Hahn, Rüdiger Wittig, and Adjima Thiombiano. 2011. "Impacts of Land-Use on West African Savanna Vegetation: A Comparison between Protected and Communal Area in Burkina Faso." *Biodiversity and Conservation* 20, no. 14 (December): 3341–62. https://doi.org/10.1007/s10531-011-0114-0.

The Nature Conservancy (TNC). 2021. "Room for Rhinos to Roam." Stories in Africa. Accessed 16 March 2021. https://www.nature.org/en-us/about-us /where-we-work/africa/stories-in-africa/relocating-rhinos-kenya/.

Ndirangu, Mwangi. 2017. "Herder Charged with Murder of Tristan Voorspuy at Sosian." *Daily Nation*, 4 April 2017. https://allafrica.com/stories/201704050049 .html.

Neely, Constance L., and Richard Hatfield. 2007. "Livestock Systems." In *Farming with Nature: The Science and Practice of Ecoagriculture*, edited by Sara Scherr and Jeffrey McNeely, 121–42. Washington: Island Press.

Nelson, Robert H. 2003. "Environmental Colonialism: 'Saving' Africa from Africans." *The Independent Review* 8, no. 1 (Summer): 65–86.

Neumann, Roderick P. 1998. *Imposing Wilderness: Struggles over Livelihood and Nature Preservation in Africa*, vol. 4. Berkeley: University of California Press.

– 2001. "Africa's 'Last Wilderness': Reordering Space for Political and Economic Control in Colonial Tanzania." *Africa* 71, no. 4 (November): 641–65. https://doi.org/10.3366/afr.2001.71.4.641.

– 2004. "Moral and Discursive Geographies in the War for Biodiversity in Africa." *Political Geography* 23, no. 7 (September): 813–37. https://doi.org/10.1016/j.polgeo.2004.05.011.

Neves, Katja Grötzner. 2019. *Postnormal Conservation: Botanic Gardens and the Reordering of Biodiversity Governance.* Albany: SUNY Press.

Nicholls, Christine S. 2005. *Red Strangers: The White Tribe of Kenya.* London: Timewell.

Njuguna, Grace Wanjiru. 2019. "Transformation of White Settler Agriculture in Colonial Kenya: The Case of Molo, Nakuru district, 1904–1963." Master's thesis, Egerton University, Nairobi.

Noe, Christine. 2019. "The Berlin Curse in Tanzania: (Re)making of the Selous World Heritage Property." *South African Geographical Journal/Suid-Afrikaanse Geografiese Tydskrif* 101, no. 3 (September): 379–98. https://doi.org/10.1080/03736245.2019.1645039.

Noor, Mohamed. 2019. "Conservation in Northern Kenya: Conflicts over Community Land in the Pastoral Margins." Future Agricultures Consortium, 13 November 2019. https://www.future-agricultures.org/blog/conservation-in-northern-kenya-conflicts-over-community-land-in-the-pastoral-margins/.

Northern Rangelands Trust (NRT). n.d. Northern Rangelands Trust Home Page. Accessed 19 October 2021. https://www.nrt-kenya.org/.

– 2013. *The Story of the Northern Rangelands Trust.* Isiolo: Northern Rangelands Trust.

– 2015. *NRT State of Conservancies Report 2014.* Isiolo: Northern Rangelands Trust.

– 2019. *The Northern Rangelands Trust Rangelands Strategy: 2019–2022.* Isiolo: Northern Rangelands Trust.

– 2020a. *Northern Rangelands Trust Status of Wildlife Report 2005–2019: Impact of NRT-Member Community Conservancies on Wildlife in Northern Kenya.* Isiolo: Northern Rangelands Trust.

– 2020b. *State of Conservancies Report 2020.* Isiolo: Northern Rangelands Trust.

– 2021. *State of Conservancies Report: 2021.* Isiolo: Northern Rangelands Trust.

Noss, Reed F., Andrew P. Dobson, Robert Baldwin, Paul Beier, Cory R. Davis, Dominick A. Dellasala, John Francis, et al. 2012. "Bolder Thinking for Conservation." *Conservation Biology* 26, no. 1 (February): 1–4. https://doi.org/10.1111/j.1523-1739.2011.01738.x.

NTV Kenya. 2017. "Public Outcry on Police Killing of Livestock in Laikipia." Uploaded 4 November 2017. YouTube video, 0:2:39. https://www.youtube.com/watch?v=eUsJ30SxwUM.

Nyaligu, Maurice O., and Susie Weeks. 2013. "An Elephant Corridor in a Fragmented Conservation Landscape: Preventing the Isolation of Mount Kenya National Park and National Reserve." *Parks* 19, no. 1 (March): 91–101. https://doi.org/10.2305/iucn.ch.2013.parks-19-1.mon.en.

Odadi, Wilfred O., Joe Fargione, and Daniel I. Rubenstein. 2017. "Vegetation, Wildlife, and Livestock Responses to Planned Grazing Management in an African Pastoral Landscape." *Land Degradation & Development* 28, no. 7 (October): 2030–8. https://doi.org/10.1002/ldr.2725.

Odukoya, Adelaja Odutola. 2018. "Settler and Non-settler Colonialism in Africa." In *The Palgrave Handbook of African Politics, Governance and Development*, edited by Samuel Ojo Oloruntoba and Toyin Falola, 173–86. New York: Palgrave Macmillan.

Ofcansky, Thomas P. 1984. "Kenya Forestry under British Colonial Administration, 1895–1963." *Journal of Forest History* 28, no. 3 (July): 136–43. https://doi.org/10.2307/4004697.

Oloruntoba, Samuel Ojo. 2020. "The Politics of Paternalism and Implications of Global Governance on Africa: A Critique of the Sustainable Development Goals." In *Pan Africanism, Regional Integration and Development in Africa*, edited by Samuel Ojo Oloruntoba, 165–79. Cham: Palgrave Macmillan.

Ol Pejeta Conservancy (OPC). n.d.-a. "Chimpanzee Sanctuary." Sweetwaters Chimpanzee Sanctuary. Accessed 23 July 2021. https://www.olpejetaconservancy.org/wildlife/chimpanzees/sweetwaters-chimpanzee-sanctuary/.

– n.d.-b. "Must See: Chimpanzees." Accessed 23 July 2021. https://www.olpejetaconservancy.org/plan-your-visit/must-see/must-see-chimpanzees/.

– n.d.-c. "Our Wonderful Wild Dogs." All News. Accessed 19 October 2021. https://www.olpejetaconservancy.org/our-wonderful-wild-dogs/.

– n.d.-d. "Sudan: A Tribute to an Icon." Accessed 23 July 2021. https://www.olpejetaconservancy.org/sudan-a-tribute-to-an-icon/.

– n.d.-e. "Support Us." Adopt a Black Rhino. Accessed 23 July 2021. https://www.olpejetaconservancy.org/adopt-a-black-rhino/black-rhinos-baraka/.

– (@OlPejetaConservancy). 2019. "Thank you for your feedback and comments…" Response to a review of Ol Pejeta Conservancy on TripAdvisor. 3 June 2019. https://en.tripadvisor.com.hk/ShowUserReviews-g10501193-d1449389-r675249911-Ol_Pejeta_Conservancy-Laikipia_County_Rift_Valley_Province.html.

– 2021a. *2021 Annual Report*. Nanyuki: Ol Pejeta Conservancy. https://www.olpejetaconservancy.org/uploads/assets/uploads/2022/08/Ol-Pejeta-Conservancy-2021-Annual-Report.pdf.

– 2021b. *Black Rhinos*. Accessed 23 July 2021. http://olpejetaconservancy.org/uploads/assets/uploads/2021/01/Ol-Pejeta-Black-Rhinos.pdf.

Olweny, Noel O., Geoffrey M. Wahungu, and Gilbert OO Obwoyere. 2020. "Browsing Impacts on Acacia drepanolobium Sjostedt and Associated Ant Guilds in Ol Pejeta Conservancy, Kenya." *Biology and Life Sciences* 62, no. 1 (October): 21. https://doi.org/10.47119/ijrp1006211020201472.

Overton, John. 1987. "The Colonial State and Spatial Differentiation: Kenya, 1895–1920." *Journal of Historical Geography* 13, no. 3 (July): 267–82. https://doi.org/10.1016/s0305-7488(87)80115-9.

Pailey, Robtel Neajai. 2021. *Development, (Dual) Citizenship and its Discontents in Africa: The Political Economy of Belonging to Liberia*, vol. 153. Cambridge: Cambridge University Press,.

Parish, Jessica., 2020. "Re-wilding Parkdale? Environmental Gentrification, Settler Colonialism, and the Reconfiguration of Nature in 21st Century Toronto." *Environment and Planning E: Nature and Space* 3, no. 1 (March): 263–86. https://doi.org/10.1177/2514848619868110.

Parreñas, Juno S. 2012. "Producing Affect: Transnational Volunteerism in a Malaysian Orangutan Rehabilitation Center." *American Ethnologist* 39, no. 4 (November): 673–87. https://doi.org/10.1111/j.1548-1425.2012.01387.x.

– 2016. "The Materiality of Intimacy in Wildlife Rehabilitation: Rethinking Ethical Capitalism through Embodied Encounters with Animals in Southeast Asia." *Positions: East Asia Cultures Critique* 24, no. 1 (February): 97–127. https://doi.org/10.1215/10679847-3320065.

– 2018. *Decolonizing Extinction: The Work of Care in Orangutan Rehabilitation*. Durham: Duke University Press.

Pasternak, Shiri, and Tia Dafnos. 2018. "How Does a Settler State Secure the Circuitry of Capital?" *Environment and Planning D: Society and Space* 36, no. 4 (August): 739–57. https://doi.org/10.1177/0263775817713209.

Patton, Felix. 2010. *Solio: The Heartbeat of Rhino Conservation for 40 Years*. Nairobi: East African Wildlife Society.

Patton, Felix, Petra Campbell, and Edward Parfet. 2007. "Establishing a Monitoring System for Black Rhinos in the Solio Game Reserve, Central Kenya." *Pachyderm* 43 (July–December): 87–95.

Patton, Felix, M.S. Mulama, Samuel. Mutisya, and P.E. Campbell. 2010a. "The Colonization of a New Area in the First Six Months Following 'Same-Day' Free Release Translocation of Black Rhinos in Kenya." *Pachyderm* 47 (January–June): 66–79.

– 2010b. "The Effect of Removing a Dividing Fence between Two Populations of Black Rhinos." *Pachyderm* 47 (January–June): 55–8.

Peake, Linda, and Eric Sheppard. 2014. "The Emergence of Radical/Critical Geography within North America." *ACME: An International Journal for Critical Geographies* 13, no. 2 (March): 305–27.

Pestalozzi, Pierre. 1986. "Historical and Present Day Agricultural Change on Mt. Kenya." In *Mount Kenya Area: Contributions to Ecology and Socio-Economy*, edited by Matthias Winiger, 105–17. Berne: University of Berne, Institute of Geography.

Pettorelli, Nathalie, Sarah M. Durant, and Johan T. Du Toit, eds. 2019. *Rewilding*. Cambridge: Cambridge University Press.

Pilcher, Helen. 2018. "Sudan the Rhino is Dead. But His Sperm Could Save the Species." *The Guardian*, 20 March 2018. https://www.theguardian.com /commentisfree/2018/mar/20/sudan-northern-white-rhino-dead-species -endangered-species-conservationists.

Prettejohn, Michael. 2012. *Endless Horizons: 100 Years of the Prettejohn Family in Kenya*. Nairobi: Old Africa Books.

Protected Planet. n.d. "Connectivity." Accessed 19 October 2021. https:// www.protectedplanet.net/en/thematic-areas/connectivity-conservation.

Pulido, Laura. 2017. "Geographies of Race and Ethnicity II: Environmental Racism, Racial Capitalism and State-Sanctioned Violence." *Progress in Human Geography* 41, no. 4 (August): 524–33. https://doi.org/10.1177 /0309132516646495.

Rainforest Foundation UK and Survival. 2020. *The "Post-2020 Global Biodiversity Framework" – A New Threat to Indigenous People and Local Communities?* London: Rainforest Foundation UK.

Ramutsindela, Maano. 2004. *Parks and People in Postcolonial Societies: Experiences in Southern Africa*. Dordrecht: Kluwer Academic Publishers.

– 2007. *Transfrontier Conservation in Africa: At the Confluence of Capital, Politics and Nature*. Wallingford, UK: CABI.

Rattan, Jasveen K., Paul F.J. Eagles, and Heather L. Mair. 2012. "Volunteer Tourism: Its Role in Creating Conservation Awareness." *Journal of Ecotourism* 11, no. 1 (March): 1–15. https://doi.org/10.1080/14724049.2011 .604129.

Render Loyalty. 2018. "Understanding Conservation: Lewa's Model." Render Loyalty website, 14 March 2018. https://www.renderloyalty.com /journal/4607/understanding-conservation-lewas-model.

RESCUE. n.d. "Who We Are." The Reteti Elephant Sanctuary. Accessed 23 July 2021. https://www.reteti.org/who-we-are.

– 2020. "Watch this space, I will outgrow this giant behind me – Lomunyak." Facebook, 28 December 2020. https://www.facebook .com/RetetiElephantSanctuary/posts/676169759669311?comment _id=676177386335215.

Reyes-García, Victoria, Álvaro Fernández-Llamazares, Yildiz Aumeeruddy-Thomas, Petra Benyei, Rainer W. Bussmann, Sara K. Diamond, David García-del-Amo, et al. 2022. "Recognizing Indigenous Peoples' and Local Communities' Rights and Agency in the Post-2020 Biodiversity Agenda." *Ambio* 51:84–92. https://doi.org/10.1007/s13280 -021-01561-7.

Riggio, Jason, Andrew P. Jacobson, Robert J. Hijmans, and Tim Caro. 2019. "How Effective Are the Protected Areas of East Africa?" *Global Ecology and Conservation* 17 (January): e00573. https://doi.org/10.1016/j.gecco.2019 .e00573.

Roffers, Frank. 2016. "Visiting the Famous Fairmont Mount Kenya Safari Club." Artful Living, 12 October 2016. https://artfulliving.com/14428-2/.

Ross, Stephen R., and Jesse G. Leinwand. 2020. "A Review of Research in Primate Sanctuaries." Biology Letters 16, no. 4 (April): 20200033. https://doi .org/10.1098/rsbl.2020.0033.

Ros-Tonen, Mirjam AF, James Reed, and Terry Sunderland. 2018. "From Synergy to Complexity: The Trend toward Integrated Value Chain and Landscape Governance." Environmental Management 62, no. 1 (July): 1–14. https://doi.org/10.1007/s00267-018-1055-0.

Saito, Natsu Taylor. 2014. "Tales of Color and Colonialism: Racial Realism and Settler Colonial Theory." Florida A&M University Law Review 10, no. 1 (Fall): 1–108.

Salazar, Joseph A. 2014. "Multicultural Settler Colonialism and Indigenous Struggle in Hawai'i: The Politics of Astronomy on Mauna a Wākea." PhD diss., University of Hawai'i at Manoa, Manoa, HI.

Saranillio, Dean Itsuji. 2013. "Why Asian Settler Colonialism Matters: A Thought Piece on Critiques, Debates, and Indigenous Difference." Settler Colonial Studies 3, nos. 3–4 (November): 280–94. https://doi.org/10.1080 /2201473X.2013.810697.

Save the Rhino. 2018. "All Rhinos Translocated to Tsavo East National Park Have Died." Save the Rhino News, 30 July 2018. https://www.savetherhino .org/africa/kenya/all-rhinos-translocated-to-tsavo-east-national-park-have -died/.

– 2020. "Scientists Have Successfully Created Northern White Rhino Embryos." Save the Rhino News, 17 January 2020. https://www .savetherhino.org/rhino-species/white-rhino/scientists-have-successfully -created-northern-white-rhino-embryos/.

Savory, Allan, with Jody Butterfield. 1999. Holistic Management: A New Framework for Decision Making. Washington: Island Press.

Sayegh, Fayez Abdullah. 1965. Zionist Colonialism in Palestine, vol. 1. Beirut: Research Center, Palestine Liberation Organization.

Schauer, Jeff. 2015. "The Elephant Problem: Science, Breaucracy, and Kenya's National Parks, 1955 to 1975." African Studies Review 58, no. 1 (April): 177–98. https://doi.org/10.1017/asr.2015.9.

Secretariat of the Convention on Biological Diversity. 2014. Global Biodiversity Outlook 4. Montréal: UNEP.

Seddon, Philip J., Christine J. Griffiths, Pritpal S. Soorae, and Doug P. Armstrong. 2014. "Reversing Defaunation: Restoring Species in a Changing World." Science 345, no. 6195 (25 July): 406–12. https://doi.org/10.1126 /science.1251818.

Sernert, Henriette. 2017. "Pilot Study of Corridor Use by African Elephants (Loxodonta africana) in Ol Pejeta Conservancy, Laikipia District, Kenya." Master's thesis, Swedish University of Agricultural Sciences, Uppsala.

Sharma, Nandita, and Cynthia Wright. 2008. "Decolonizing Resistance, Challenging Colonial States." *Social Justice* 35, no. 3 (113): 120–38.

Sheldrick, Daphne. 2012. *Love, Life, and Elephants: An African Love Story*. New York: Farrar, Straus and Giroux.

Sheldrick Wildlife Trust (SWT). n.d.-a. "Aerial Surveillance." Accessed 23 July 2021. https://www.sheldrickwildlifetrust.org/projects/aerial-surveillance.

– n.d.-b. "Lenana." Accessed 22 July 2021. https://www.sheldrickwildlifetrust.org/orphans/lenana.

– n.d.-c. "Raising an Orphan." Accessed 23 July 2021. https://www.sheldrickwildlifetrust.org/projects/orphans/raise-an-orphan-elephant.

– n.d.-d. "The Rescue of Lenana." Accessed 22 July 2021. https://www.sheldrickwildlifetrust.org/news/updates/the-rescue-of-lenana.

– 2019. "In Memoriam – Dame Daphne One Year On." 12 April 2019. https://www.sheldrickwildlifetrust.org/news/updates/in-memoriam-dame-daphne-one-year-on.

Sheldrick Wildlife Trust USA. 2018. *Sheldrick Wildlife Trust USA: 2018 Annual Report to Contributors*. Laguna Hills, CA: Sheldrick Wildlife Trust USA.

Shukin, Nicole. 2011. "Transfections of Animal Touch, Techniques of Biosecurity." *Social Semiotics* 21, no. 4 (September): 483–501. https://doi.org/10.1080/10350330.2011.591994.

Sindiga, Isaac. 1995. "Wildlife Based Tourism in Kenya: Land Use Conflicts and Government Compensation Policies over Protected Areas." *Journal of Tourism Studies* 6, no. 2 (December): 45–55.

Singh, Jaidev, and Henk Van Houtum. 2002. "Post-colonial Nature Conservation in Southern Africa: Same Emperors, New Clothes?" *GeoJournal* 58, no. 4 (December): 253–63. https://doi.org/10.1023/b:gejo.0000017956.82651.41.

Srinivasan, Krithika, and Rajesh Kasturirangan. 2016. "Political Ecology, Development, and Human Exceptionalism." *Geoforum* 75 (October): 125–8. https://doi.org/10.1016/j.geoforum.2016.07.011.

Stears, Keenan, Melissa H. Schmitt, Wendy C. Turner, Douglas J. McCauley, Epaphras A. Muse, Halima Kiwango, Daniel Mathayo, and Benezeth M. Mutayoba. 2021. "Hippopotamus Movements Structure the Spatiotemporal Dynamics of an Active Anthrax Outbreak." *Ecosphere* 12, no. 6 (June): e03540. https://doi.org/10.1002/ecs2.3540.

Steinhart, Edward. 1989. "Hunters, Poachers and Gamekeepers: Towards a Social History of Hunting in Colonial Kenya." *Journal of African History* 30, no. 2 (July): 247–64. https://doi.org/10.1017/s0021853700024129.

– 2006. *Black Poachers, White Hunters: A Social History of Hunting in Colonial Kenya*. Athens: Ohio University Press.

Strum, Shirley C., Graham Stirling, and Steve Kalusi Mutunga. 2015. "The Perfect Storm: Land Use Change Promotes *Opuntia stricta*'s Invasion of Pastoral Rangelands in Kenya." *Journal of Arid Environments* 118 (July): 37–47. https://doi.org/10.1016/j.jaridenv.2015.02.015.

Sundaresan, Siva R., and Corinna Riginos. 2010. "Lessons Learned from Biodiversity Conservation in the Private Lands of Laikipia, Kenya." *Great Plains Research* 20, no. 1 (Spring): 17–27.

Sutherland, Klinette, Mduduzi Ndlovu, and Antón Pérez-Rodríguez. 2018. "Use of Artificial Waterholes by Animals in the Southern Region of the Kruger National Park, South Africa." *African Journal of Wildlife Research* 48, no. 2 (October): 1–14. https://doi.org/10.3957/056.048.023003.

Suzuki, Yuka. 2001. "Drifting Rhinos and Fluid Properties: The Turn to Wildlife Production in Western Zimbabwe." *Journal of Agrarian Change* 1, no. 4 (October): 600–25. https://doi.org/10.1111/1471-0366.00020.

Szapary, Peter. 2000. "The Lewa Wildlife Conservancy in Kenya: A Case Study." In *Wildlife Conservation by Sustainable Use*, edited by Herbert H.T. Prins, Jan Geu Grootenhuis, and Thomas T. Dolan, 35–50. Boston: Kluwer Academic Publishers.

Sze, Jocelyne S., L. Roman Carrasco, Dylan Childs, and David P. Edwards. 2022. "Reduced Deforestation and Degradation in Indigenous Lands Pan-tropically." *Nature Sustainability* 5, no. 2 (February): 123–30. https://doi.org/10.1038/s41893-021-00815-2.

Szott, Isabelle D., Yolanda Pretorius, and Nicola F. Koyama. 2019. "Behavioural Changes in African Elephants in Response to Wildlife Tourism." *Journal of Zoology* 308, no. 3 (July): 164–74. https://doi.org/10.1111/jzo.12661.

Taschereau Mamers, Danielle. 2019. "Human-Bison Relations as Sites of Settler Colonial Violence and Decolonial Resurgence." *Humanimalia* 10, no. 2 (February): 10–41. https://doi.org/10.52537/humanimalia.9500.

– 2020. "'Last of the Buffalo': Bison Extermination, Early Conservation, and Visual Records of Settler Colonization in the North American West." *Settler Colonial Studies* 10, no. 1 (January): 126–47. https://doi.org/10.1080/2201473x.2019.1677134.

Taylor, Madyson, Chris E. Hurst, Michela J. Stinson, and Bryan S.R. Grimwood. 2020. "*Becoming Care-full*: Contextualizing Moral Development among Captive Elephant Volunteer Tourists to Thailand." *Journal of Ecotourism* 19, no. 2 (April): 113–31. https://doi.org/10.1080/14724049.2019.1657125.

Thakholi, Lerato. 2021. "Conservation Labour Geographies: Subsuming Regional Labour into Private Conservation Spaces in South Africa." *Geoforum* 123:1–11. https://doi.org/10.1016/j.geoforum.2021.04.019.

Thirgood, Simon, Anna Mosser, Sebastian Tham, Grant Hopcraft, Ephraim Mwangomo, Titus Mlengeya, Morris Kilewo, John Fryxell, A.R.E. Sinclair, and Markus Borner. 2004. "Can Parks Protect Migratory Ungulates? The Case of the Serengeti Wildebeest." *Animal Conservation Forum* 7, no. 2 (May): 113–20. https://doi.org/10.1017/s1367943004001404.

Thomas, Neil William. 2000. "From Herdsmen to Safari Guides: An Assessment of Environmental Partnerships at IL Ngwesi, Laikipia District, Kenya." Master's thesis, University of Natal, Durban, South Africa.

Thomson, Joseph. 1887. *Through Masai Land: A Journey of Exploration among the Snowclad Volcanic Mountains and Strange Tribes of Eastern Equatorial Africa*. London: Gilbert and Rivington.

Thouless, C.R., and J. Sakwa. 1995. "Shocking Elephants: Fences and Crop Raiders in Laikipia District, Kenya." *Biological conservation* 72 (1): 99–107. https://doi.org/10.1016/0006-3207(94)00071-w.

Todd, Zoe. 2014. "Fish Pluralities: Human-Animal Relations and Sites of Engagement in Paulatuuq, Arctic Canada." *Études Inuit Studies* 38, nos. 1–2 (February): 217–38. https://doi.org/10.7202/1028861ar.

– 2017. "Commentary: The Environmental Anthropology of Settler Colonialism, Part I." *Engagement* (blog). The Anthropology and Environment Society, 17 April 2017. https://aesengagement.wordpress.com/2017/04/11/commentary-the-environmental-anthropology-of-settler-colonialism-part-i/.

– 2022. "Fossil Fuels and Fossil Kin: An Environmental Kin Study of Weaponised Fossil Kin and Alberta's So-Called 'Energy Resources Heritage.'" *Antipode*. Published in Early View, 8 November 2022. https://doi.org/10.1111/anti.12897.

Tomiak, Julie. 2016. "Unsettling Ottawa: Settler Colonialism, Indigenous Resistance, and the Politics of Scale." *Canadian Journal of Urban Research* 25, no. 1 (Summer): 8–21.

Trzebinski, Errol. 1986. *The Kenya Pioneers*. New York: W.W. Norton.

Tuck, Eve, and K. Wayne Yang. 2012. Decolonization is Not a Metaphor. *Decolonization: Indigeneity, Education & Society* 1, no. 1 (September): 1–40.

– 2014. "Unbecoming Claims: Pedagogies of Refusal in Qualitative Research." *Qualitative Inquiry* 20, no. 6 (July): 811–18. https://doi.org/10.1177/1077800414530265.

Turnhout, Esther, and Andy Purvis. 2020. "Biodiversity and Species Extinction: Categorisation, Calculation, and Communication." *Griffith Law Review* 29, no. 4 (October): 669–85. https://doi.org/10.1080/10383441.2020.1925204.

UN Environment Programme World Conservation Monitoring Centre (UNEP-WCMC), International Union for Conservation of Nature (IUCN), and National Geographic Society (NGS). 2018. *Protected Planet Report 2018*. Cambridge, Gland, and Washington: United Nations Environment Programme.

United Nations Development Programme (UNDP). 2003. "Repatriation of the Mountain Bongo Antelope to Mt. Kenya World Heritage Site." GEF Small

Grants Programme, 1 November 2003. https://sgp.undp.org/resources-155/our-stories/219-repatriation-of-the-mountain-bongo-antelope-to-mt-kenya-world-heritage-site.html.

United Nations Environment Program (UNEP). 2020. *Improving Ways of Addressing Connectivity in the Conservation of Migratory Species.* UNEP/CMS/Resolution 12.26 (Rev.COP13), February 2020. https://www.cms.int/sites/default/files/document/cms_cop13_res.12.26_rev.cop13_e.pdf.

Upadhyay, Nishant. 2019. "Making of 'Model' South Asians on the Tar Sands: Intersections of Race, Caste, and Indigeneity." *Critical Ethnic Studies* 5, nos. 1–2 (Spring): 152–73. https://doi.org/10.5749/jcritethnstud.5.1-2.0152.

Van den Akker, M.L. 2016. "Monument of Nature? An Ethnography of the World Heritage of Mt. Kenya." Unpublished PhD diss., Leiden University, Leiden, Netherlands.

Velednitsky, Stephan, Sara N.S. Hughes, and Rhys Machold. 2020. "Political Geographical Perspectives on Settler Colonialism." *Geography Compass* 14, no. 6 (June): e12490. https://doi.org/10.1111/gec3.12490.

Veracini, Lorenzo. 2008. "Colonialism and Genocides: Notes for the Analysis of the Settler Archive." In *Empire, Colony, Genocide: Conquest, Occupation, and Subaltern Resistance in World History*, edited by Dirk A. Moses, 148–61. Oxford: Berghahn Books.

– 2010. *Settler Colonialism: A Theoretical Overview.* Houndmills, UK: Palgrave Macmillan.

Verweijen, Judith, and Esther Marijnen. 2018. "The Counterinsurgency/Conservation Nexus: Guerrilla Livelihoods and the Dynamics of Conflict and Violence in the Virunga National Park, Democratic Republic of the Congo." *Journal of Peasant Studies* 45, no. 2 (February): 300–20. https://doi.org/10.1080/03066150.2016.1203307.

Wa-Githumo, Mwangi. 1991. "The Truth about the Mau Mau Movement: The Most Popular Uprising in Kenya." *Transafrican Journal of History* 20:1–18.

Waithaka, John. 2012. "Historical Factors that Shaped Wildlife Conservation in Kenya." *The George Wright Forum* 29 (1): 21–9.

Waller, Richard. 2004. "'Clean' and 'Dirty': Cattle Disease and Control Policy in Colonial Kenya, 1900–40." *The Journal of African History* 45, no. 1 (March): 45–80. https://doi.org/10.1017/s0021853703008508.

Wargute, Patrick W. 2007. *Numbers and Distribution of Wildlife in Kwale, Kilifi, Malindi, and Lamu Districts: Increased Data for Conservation and Management (1977–2006).* No. 170. Nairobi: Department of Resource Surveys and Remote Sensing, Ministry of Environment and Natural Resources.

Watson, Rupert. 2014. *Culture Clash: The Death of a District Commissioner in the Loita Hills.* Naivasha, Kenya: Old Africa Books.

Weldemichel, Teklehaymanot G. 2020. "Othering Pastoralists, State Violence, and the Remaking of Boundaries in Tanzania's Militarised Wildlife Conservation Sector." *Antipode* 52, no. 5 (September): 1496–518. https://doi.org/10.1111/anti.12638.

West, T.F. 1959. "The History of the African Pyrethrum Industry." *Journal of the Royal Society of Arts* 107, no. 5034 (May): 423–41.

Western, David, and Helen Gichohi. 1993. "Segregation Effects and the Impoverishment of Savanna Parks: The Case for Ecosystem Viability Analysis." *African Journal of Ecology* 31, no. 4 (December): 269–81. https://doi.org/10.1111/j.1365-2028.1993.tb00541.x.

Western, David, Samantha Russell, and Innes Cuthill. 2009. "The Status of Wildlife in Protected Areas Compared to Non-protected Areas of Kenya." *PloS one* 4, no. 7 (8 July): e6140. https://doi.org/10.1371/journal.pone.0006140.

Whyte, Kyle. 2018a. "Settler Colonialism, Ecology, and Environmental Injustice." *Environment and Society* 9, no. 1 (September): 125–44. https://doi.org/10.3167/ares.2018.090109.

– 2018b. "What Do Indigenous Knowledges Do for Indigenous Peoples?" In *Traditional Ecological Knowledge: Learning from Indigenous Practices for Environmental Sustainability*, edited by Melissa K. Nelson and Daniel Shilling, 1–20. Cambridge: Cambridge University Press.

WildLandscapes International. n.d. *Ukanda wa Vifaru: Securing a Landscape-Scale Rhino Corridor in Kenya*. Buckingham: WildLandscapes International and Ol Pejeta Conservancy.

Wilson, Edward O. 2016. *Half-Earth: Our Planet's Fight for Life*. New York: W.W. Norton.

Winmill, Natalie Elizabeth. 2014. "The Mount Kenya Elephant Corridor's Utilisation by Elephants and Its Perceived Success." Master's thesis, University of Southampton, Southampton.

Winter, Stuart. 2017. "Last Northern Male White Rhino Joins TINDER to Save his Species: 'I'm One of a Kind.'" *Express*, 26 April 2017. https://www.express.co.uk/news/nature/796819/Rhino-on-Tinder- Sudan-last-white-endangered-species-creatures-Kenya.

Witt, Arne B.R., Winnie Nunda, Tim Beale, and Darren J. Kriticos. 2020. "A Preliminary Assessment of the Presence and Distribution of Invasive and Potentially Invasive Alien Plant Species in Laikipia County, Kenya, a Biodiversity Hotspot." *Koedoe: African Protected Area Conservation and Science* 62 (2): a1605. https://doi.org/10.4102/koedoe.v62i1.1605.

Witucki, Lawrence A. 1976. *Agricultural Development in Kenya Since 1967*. No. 123. Washington: US Department of Agriculture, Economic Research Service [Foreign Demand and Competition Division].

Wolfe, Patrick. 1999. *Settler Colonialism*. London: A&C Black.

– 2006. "Settler Colonialism and the Elimination of the Native." *Journal of Genocide Research* 8, no. 4 (December): 387–409. https://doi.org/10.1080/14623520601056240.

Wuerthner, George, Eileen Crist, and Tom Butler, eds. 2015. *Protecting the Wild: Parks and Wilderness, the Foundation for Conservation*. Washington: Island Press.

Ybarra, Megan. 2018. *Green Wars: Conservation and Decolonization in the Maya Forest*. Berkeley: University of California Press.

Index